Density Waves in Solids

Density Waves in Solids

George Grüner

Department of Physics
and Solid State Science Center
University of California, Los Angeles

CRC Press is an imprint of the
Taylor & Francis Group, an **informa** business

The Advanced Book Program

First published 1994 by Westview Press

Published 2018 by CRC Press
Taylor & Francis Group
6000 Broken Sound Parkway NW, Suite 300
Boca Raton, FL 33487-2742

Visit the Taylor & Francis Web site at
http://www.taylorandfrancis.com

and the CRC Press Web site at
http://www.crcpress.com

Library of Congress Cataloging-in-Publication Data
Grüner, George.
Density waves in solids/George Grüner.
p. cm.—(Frontiers in physics ; v. 89)
Includes bibliographical references and index.
ISBN 0-201-62654-3
1. Charge density waves. 2. Energy-band theory of solids.
I. Title. II. Series.
QC176.8.E4G78 1994
530.4'16—dc20 93-32362
CIP

ISBN: 0-7382-0304-1

ISBN 13: 978-0-7382-0304-1 (pbk)

Text design by Joyce Weston
Set in 10.5 point Palatino by Science Typographers

Frontiers in Physics

David Pines, Editor

Volumes of the Series published from 1961 to 1973 are not officially numbered. The parenthetical numbers shown are designed to aid librarians and bibliographers to check the completeness of their holdings.

Titles published in this series prior to 1987 appear under either the W. A. Benjamin or the Benjamin/Cummings imprint; titles published since 1986 appear under the Addison-Wesley imprint.

1.	N. Bloembergen	Nuclear Magnetic Relaxation: A Reprint Volume, 1961
2.	G. F. Chew	S-Matrix Theory of Strong Interactions: A Lecture Note and Reprint Volume, 1961
3.	R. P. Feynman	Quantum Electrodynamics: A Lecture Note and Reprint Volume
4.	R. P. Feynman	The Theory of Fundamental Processes: A Lecture Note Volume, 1961
5.	L. Van Hove, N. M. Hugenholtz L. P. Howland	Problem in Quantum Theory of Many-Particle Systems: A Lecture Note and Reprint Volume, 1961
6.	D. Pines	The Many-Body Problem: A Lecture Note and Reprint Volume, 1961
7.	H. Frauenfelder	The Mössbauer Effect: A Review—With a Collection of Reprints, 1962
8.	L. P. Kadanoff G. Baym	Quantum Statistical Mechanics: Green's Function Methods in Equilibrium and Nonequilibrium Problems, 1962
9.	G. E. Pake	Paramagnetic Resonance: An Introductory Monograph, 1962 [cr. (42)—2nd edition]G. E. Pake
10.	P. W. Anderson	Concepts in Solids: Lectures on the Theory of Solids, 1963
11.	S. C. Frautschi	Regge Poles and S-Matrix Theory, 1963
12.	R. Hofstadter	Electron Scattering and Nuclear and Nucleon Structure: A Collection of Reprints with an Introduction, 1963
13.	A. M. Lane	Nuclear Theory: Pairing Force Correlations to Collective Motion, 1964
14.	R. Omnès M. Froissart	Mandelstam Theory and Regge Poles: An Introduction for Experimentalists, 1963
15.	E. J. Squires	Complex Angular Momenta and Particle Physics: A Lecture Note and Reprint Volume, 1963
16.	H. L. Frisch J. L. Lebowitz	The Equilibrium Theory of Classical Fluids: A Lecture Note and Reprint Volume, 1964

	Author(s)	Title
17.	M. Gell-Mann Y. Ne'eman	The Eightfold Way (A Review—With a Collection of Reprints), 1964
18.	M. Jacob G. F. Chew	Strong-Interaction Physics: A Lecture Note Volume, 1964
19.	P. Nozières	Theory of Interacting Fermi Systems, 1964
20.	J. R. Schrieffer	Theory of Superconductivity, 1964 (revised 3rd printing, 1983)
21.	N. Bloembergen	Nonlinear Optics: A Lecture Note and Reprint Volume, 1965
22.	R. Brout	Phase Transitions, 1965
23.	I. M. Khalatnikov	An Introduction to the Theory of Superfluidity, 1965
24.	P. G. deGennes	Superconductivity of Metals and Alloys, 1966
25.	W. A. Harrison	Pseudopotentials in the Theory of Metals, 1966
26.	V. Barger D. Cline	Phenomenological Theories of High Energy Scattering: An Experimental Evaluation, 1967
27.	P. Choquàrd	The Anharmonic Crystal, 1967
28.	T. Loucks	Augmented Plane Wave Method: A Guide to Performing Electronic Structure Calculations—A Lecture Note and Reprint Volume, 1967
29.	Y. Ne'eman	Algebraic Theory of Particle Physics: Hadron Dynamics In Terms of Unitary Spin Current, 1967
30.	S. L. Adler R. F. Dashen	Current Algebras and Applications to Particle Physics, 1968
31.	A. B. Migdal	Nuclear Theory: The Quasiparticle Method, 1968
32.	J. J. J. Kokkedee	The Quark Model, 1969
33.	A. B. Migdal	Approximation Methods in Quantum Mechanics, 1969
34.	R. Z. Sagdeev	Nonlinear Plasma Theory, 1969
35.	J. Schwinger	Quantum Kinematics and Dynamics, 1970
36.	R. P. Feynman	Statistical Mechanics: A Set of Lectures, 1972
37.	R. P. Feynman	Photon-Hadron Interactions, 1972
38.	E. R. Caianiello	Combinatorics and Renormalization in Quantum Field Theory, 1973
39.	G. B. Field H. Arp J. N. Bahcall	The Redshift Controversy, 1973
40.	D. Horn F. Zachariasen	Hadron Physics at Very High Energies, 1973
41.	S. Ichimaru	Basic Principles of Plasma Physics: A Statistical Approach, 1973 (2nd printing, with revisions, 1980)
42.	G. E. Pake T. L. Estle	The Physical Principles of Electron Paramagnetic Resonance, 2nd Edition, completely revised, enlarged, and reset, 1973 [cf. (9)—1st edition

Volumes published from 1974 onward are being numbered as an integral part of the bibliography.

	Author(s)	Title
43.	R. C. Davidson	Theory of Nonneutral Plasmas, 1974

44.	S. Doniach E. H. Sondheimer	Green's Functions for Solid State Physicists, 1974
45.	P. H. Frampton	Dual Resonance Models, 1974
46.	S. K. Ma	Modern Theory of Critical Phenomena, 1976
47.	D. Forster	Hydrodynamic Fluctuation, Broken Symmetry, and Correlation Functions, 1975
48.	A. B. Migdal	Qualitative Methods in Quantum Theory, 1977
49.	S. W. Lovesey	Condensed Matter Physics: Dynamic Correlations, 1980
50.	L. D. Faddeev A. A. Slavnov	Gauge Fields: Introduction to Quantum Theory, 1980
51.	P. Ramond	Field Theory: A Modern Primer, 1981 [cf. 74—2nd ed.]
52.	R. A. Broglia A. Winther	Heavy Ion Reactions: Lecture Notes Vol. I, Elastic and Inelastic Reactions, 1981
53.	R. A. Broglia A. Winther	Heavy Ion Reactions: Lecture Notes Vol. II, 1990
54.	H. Georgi	**Lie Algebras in Particle Physics, 2nd edition, 1999**
55.	P. W. Anderson	Basic Notions of Condensed Matter Physics, 1983
56.	C. Quigg	Gauge Theories of the Strong, Weak, and Electromagnetic Interactions, 1983
57.	S. I. Pekar	Crystal Optics and Additional Light Waves, 1983
58.	S. J. Gates M. T. Grisaru M. Rocek W. Siegel	Superspace *or* One Thousand and One Lessons in Supersymmetry, 1983
59.	R. N. Cahn	Semi-Simple Lie Algebras and Their Representations, 1984
60.	G. G. Ross	Grand Unified Theories, 1984
61.	S. W. Lovesey	Condensed Matter Physics: Dynamic Correlations, 2nd Edition, 1986
62.	P. H. Frampton	Gauge Field Theories, 1986
63.	J. I. Katz	High Energy Astrophysics, 1987
64.	T. J. Ferbel	Experimental Techniques in High Energy Physics, 1987
65.	T. Appelquist A. Chodos P. G. O. Freund	Modern Kaluza-Klein Theories, 1987
66.	G. Parisi	Statistical Field Theory, 1988
67.	R. C. Richardson E. N. Smith	Techniques in Low-Temperature Condensed Matter Physics, 1988
68.	J. W. Negele H. Orland	Quantum Many-Particle Systems, 1987
69.	E. W. Kolb M. S. Turner	The Early Universe, 1990
70.	E. W. Kolb M. S. Turner	The Early Universe: Reprints, 1988
71.	V. Barger R. J. N. Phillips	Collider Physics, 1987
72.	T. Tajima	Computational Plasma Physics, 1989

73. W. Kruer — The Physics of Laser Plasma Interactions, 1988
74. P. Ramond — Field Theory: A Modern Primer, 2nd edition, 1989 [cf. 51—1st edition]
75. B. F. Hatfield — Quantum Field Theory of Point Particles and Strings, 1989
76. P. Sokolsky — Introduction to Ultrahigh Energy Cosmic Ray Physics, 1989
77. R. Field — Applications of Perturbative QCD, 1989
80. J. F. Gunion, H. E. Haber, G. Kane, S. Dawson — The Higgs Hunter's Guide, 1990
81. R. C. Davidson — Physics of Nonneutral Plasmas, 1990
82. E. Fradkin — Field Theories of Condensed Matter Systems, 1991
83. L. D. Faddeev, A. A. Slavnov — Gauge Fields, 1990
84. R. Broglia, A. Winther — Heavy Ion Reactions, Parts I and II, 1990
85. N. Goldenfeld — Lectures on Phase Transitions and the Renormalization Group, 1992
86. R. D. Hazeltine, J. D. Meiss — Plasma Confinement, 1992
87. S. Ichimaru — Statistical Plasma Physics, Volume I: Basic Principles, 1992
88. S. Ichimaru — Statistical Plasma Physics, Volume II: Condensed Plasmas, 1994
89. G. Grüner — Density Waves in Solids, 1994
90. S. Safran — Statistical Thermodynamics of Surfaces, Interfaces, and Membranes, 1994
91. B. d'Espagnat — Veiled Reality: An Analysis of Present Day Quantum Mechanical Concepts, 1994
92. J. Bahcall, R. Davis, Jr., P. Parker, A. Smirnov, R. Ulrich — Solar Neutrinos: The First Thirty Years
93. R. Feynman, F. Morinigo, W. Wagner — Feynman Lectures on Gravitation
94. M. Peskin, D. Schroeder — An Introduction to Quantum Field Theory
95. R. Feynman — Feynman Lectures on Computation
96. M. Brack, R. Bhaduri — Semiclassical Physics
97. D. Cline — Weak Neutral Currents
98. T. Tajima, K. Shibata — Plasma Astrophysics
99. J. Rammer — Quantum Transport Theory
100. R. Hazeltine, F. Waelbroeck — The Framework of Plasma Physics
101. P. Ramond — Journeys Beyond the Standard Model

Editor's Foreword

The problem of communicating recent developments in a coherent fashion in the most exciting and active fields of physics continues to be with us. The enormous growth in the number of physicists has tended to make the familiar channels of communication considerably less effective. It has become increasingly difficult for experts in a given field to keep up with the current literature; the novice can only be confused. What is needed is both a consistent account of a field and the presentation of a definite "point of view" concerning it. Formal monographs cannot meet such a need in a rapidly developing field, while the review article seems to have fallen into disfavor. Indeed, it would seem that the people who are most actively engaged in developing a given field are the people least likely to write at length about it.

Frontiers in Physics was conceived in 1961 in an effort to improve the situation in several ways. Leading physicists frequently give a series of lectures, a graduate seminar, or a graduate course in their special fields of interest. Such lectures serve to summarize the present status of a rapidly developing field and may well constitute the only coherent account available at the time. One of the principal purposes of the *Frontiers in Physics* series is to make notes on such lectures available to the wider physics community.

As *Frontiers in Physics* has evolved, a second category of book, the informal text/monograph, an intermediate step between lecture notes and formal texts or monographs, has played an increasingly important role in the series. In an informal text or monograph an author has reworked his or her lecture notes to the point at which the manuscript represents a coherent summation of a newly developed field, complete with references and problems, suitable for either classroom teaching or individual study.

During the past decade the study of charge and spin density waves in highly anisotropic solids has provided a striking example of the influence of electron-electron and electron-phonon interactions in determining system behavior. Through his seminal experiments and his careful attention to comparing theory with experiment, George Grüner has played a leading role in elucidating that behavior. In this lecture-note volume, intended for a graduate student and advanced undergraduate student audience, he provides a lucid introduction to this important frontier topic in condensed matter physics. It gives me great pleasure to welcome him to the ranks of authors represented in "Frontiers in Physics."

Contents

Notation Legend

N_0	number of electrons
L	length of chain
N	number of lattice sites per unit length and per spin direction
N_e	number of electrons per unit length
n_0	number of electrons per unit volume, $N_e n_\perp$
$n_\perp$	number of chains per unit cross-sectional area
n	number of electrons per unit cell$/cm^3$
$n(\epsilon)$	density of states per spin direction
ρ_0	charge density per lattice constant
$\rho^{\mathrm{ind}}(\vec{r})$	induced charge density
ϵ_F	Fermi energy
k_F	Fermi wavevector
v_F	Fermi velocity
m_e	free electron mass
$f(\epsilon_k)$	Fermi distribution function
$d_\parallel, a$	lattice constant along the chain direction x
$b, c, d_\perp$	lattice constant perpendicular to chain direction y, z
t	transfer integral
m_b	bandmass
χ_p	Pauli susceptibility
ω_p	plasma frequency
g	gyromagnetic factor
χ_0	susceptibility (unrenormalized)
σ	conductivity
τ	relaxation time
$u(x)$	lattice displacement
C	specific heat
$\chi(q)$	Lindhard response functions
S(q)	correlation function

I(q)	scattering intensity
λ_0	period of density wave
λ	electron-phonon coupling constant (dimensionless)
λ_e	electron-electron coupling constant (dimensionless)
g_q	coupling constant
Δ	order parameter ($= \lvert\Delta\rvert e^{i\phi}$)
ω_{2k_F}	unrenormalized phonon frequency
$\omega_{\text{ren},2k_F}$	renormalized phonon frequency
V	lattice potential
M	ionic mass
Q_q, P_q	normal coordinates and conjugate momenta
$a_k^\dagger(a_k)$	electron creation (annihilation) operators
$b_q^\dagger(b_q)$	phonon creation (annihilation) operators
ρ_q	q^{th} component of electron density ($= \Sigma a_{k+q}^\dagger a_k$)
$T_{\text{CDW}}^{\text{MF}}$	mean field CDW transition temperature
U_k, V_k	transformation parameters
$\gamma_k(\gamma_k^+)$	transformed states
γ_e	electronic specific heat coefficient
f	temperature dependent condensate density
a, b, c, d	Ginzburg-Landau parameters
F	free energy
E	energy (total, lattice, condensation...)
H	Miller index of reflection
U	Coulomb interaction energy
$\vec{H}$	magnetic field
$\vec{M}$	magnetization
$T_{\text{SDW}}^{\text{MF}}$	mean field SDW transition temperature
μ	magnetic moment
μ_B	Bohr magneton
$\vec{S}$	electron spin
J_{eff}	exchange coupling constant
D^*, E^*	magnetic anisotropy constants
$\langle a_0 \rangle$	hyperfine interaction constant
H_{sf}	spin flop field
$\xi_{1D}, \xi_{\parallel}$	one-dimensional coherence length
T^*	$1D - 3D$ crossover temperature
ω_A, ω_ϕ	amplitude and phase oscillation frequencies
$\delta(x, t)$	amplitude fluctuation of order parameter
$\phi(x, t)$	phase fluctuation of order parameter

$\mathscr{L}$	Lagrangian
v_d	drift velocity
I	electric current
j	electric current density
ω	angular frequency of ionic oscillations
m^*	effective mass
c_ϕ	phason velocity
P	polarization
ϵ_0	background dielectric constant
q_0	Thomas-Fermi wave vector
$\Pi(x)$	momentum density
n_m	number of modes per unit length
C_v^ϕ	phason specific heat (Debye approx.)
Θ_D^ϕ	phason Debye temperature
Γ	damping constant of collective mode
$\beta(\omega)$	Bose-Einstein factor
$I(\omega)$	Raman scattering intensity
e^*	effective charge of solitons
l_s	spatial separation of solitons
d	spacial extension of solitons
$\vec{R}_i$	impurity site
$V_{\text{imp}}(\phi)$	impurity potential
L_0	phase-phase correlation length
j_{DW}	collective mode current density
E	electric field
$\sigma_{\text{coll}}(k, \omega)$	collective mode conductivity
$\sigma_{\text{sp}}(\omega)$	single particle conductivity
ω_g	gap frequency
$f(\omega)$	oscillator strength
l	mean free path
$\sigma_{\text{m}}(\omega)$	metallic state conductivity
$I_{\text{coll}}(\omega)$	collective mode spectral weight
$I_{\text{sp}}(\omega)$	single particle spectral weight
ϵ	dielectric constant
ω_P^*	collective mode plasma frequency
$P(\vec{r})$	polarization
ω_n	optical phonon frequencies
λ_n	dimensionless electron-phonon coupling constant
$D_\phi(\omega)$	propagator
E_T	threshold field

σ_c	cordial conductivity
σ_d	differential conductivity
$\langle j \rangle$	time averaged current density per chain
f_0	current oscillation frequency
I_n	spectral density of harmonics
C_Δ, C_ϕ	elastic constants related to change of amplitude and phase of order parameter

Preface

Density waves are broken symmetry states of metals, brought about by electron-phonon or by electron-electron interactions. The ground states are the coherent superposition of electron-hole pairs, and, as the name implies, the charge density or spin density is not uniform but displays a periodic spatial variation. The former is called the charge density wave (CDW), the latter the spin density wave (SDW) state of metals.

Charge density waves were first discussed by Fröhlich in 1954 and by Peierls in 1955; spin density wave states were postulated by Overhauser in 1962. It was recognized early that highly anisotropic band structures are important in leading to these ground states. Not surprisingly, experimental evidence for these ground states was found much later, when materials with a linear chain structure and metallic properties were discovered and investigated. Several groups of both organic and inorganic materials are now standard examples of density wave ground states; some members of these groups have been investigated in detail by a wide array of experimental techniques.

These notes give a fairly elementary, unsophisticated, and sometimes oversimplified discussion of the field. They reflect an experimentalist view; there is an emphasis on the close relation between theory and experiment—an important aspect of the field. The notes are based on lectures I gave at the Eidgenössische Technische Hochschule, Zurich in 1989 and subsequently at the University of California, Los Angeles in 1991; in both occasions to an audience including graduate and undergraduate students.

Because density waves arise in their simplest form in highly anisotropic (so-called quasi-one-dimensional materials) some fundamental aspects of the low dimensional electron gas will be discussed first. Chapter 2 focuses on the materials, on the various groups of so-called linear chain compounds. This is followed by a discussion of the mean field theory of CDW and SDW ground states and the basic experimental observations in Chapters 3 and 4. Because of the low dimensionality, fluctuation effects are important, and the phase transition is different from what is predicted by mean field theory. The nature of the phase transitions and fluctuations are discussed in Chapter 5; followed in Chapter 6 by a survey of the collective excitations called phasons, amplitudons, and magnons. Chapter 7 deals with the interaction between the ground states and the underlying lattice, and Chapter 8 with the interaction between density waves and impurities. This is followed by a discussion of electrodynamics and nonlinear transport phenomena in Chapters 9 and 10. One of the most spectacular observations in the field is the detection of current oscillations, and various interference phenomena which occur when both *dc* and *ac* driving fields are amplified. A recent review, adopted in part from *Progress in Low Temperature Physics* concludes these notes.

The phase transitions and essential features of the ground states are discussed by using second quantization formalism. While the various density wave states, together with the superconducting ground state, can be discussed using a simple Hamiltonian with a q dependent interaction potential $V(q)$ between the electrons, a traditional approach will be followed here: the charge density wave state is described starting from the Fröhlich Hamiltonian of electron-phonon interactions, while the spin density wave state will be discussed by treating the electron-electron interactions within the framework of the Hubbard model. Fluctuation effects and elementary excitations will be described within the framework of Ginzburg-Landau theory, and the interaction between density waves and the underlying lattice together with density wave-impurity interactions will also be discussed using this approach.

Several aspects of the ground states, phase transitions, and various excitations are similar to those of the superconducting state, and consequently extensive use will be made of expressions which have been worked out for BCS superconductors. These expressions will not be derived, but will merely be adopted from the literature. The same applies for the discussion of magnetic excitations which occur in the spin density wave state; which are similar to spin wave excitations well known for antiferromagnets.

Various topics, such as the microscopic description of the interaction of the collective modes with impurities, or some aspects of nonlinear transport, require a discussion which goes beyond the framework of these notes. In these cases only a short summary of the pertinent results will be given.

The field is relatively new, and is by no means a closed chapter of solid state physics. Consequently, many of the issues have not been completely resolved (this is particularly true for spin density waves) and parts of these notes reflect this "unfinished" aspect of the field. Several topics will not be covered by these notes. Density waves which arise in higher dimensions, such as the two-dimensional charge density waves observed in a certain group of materials, called dichalcogenides, and spin density waves in chromium and in its alloys, lie outside the scope of these notes. Also, the focus is on the simplest case of density waves: on the periodic modulations of the charge or spin density with a period which is incommensurate to the underlying lattice. Somewhat more complicated density waves, which arise in materials which have two conducting chains—with the material tetrathiafulvalene-tetracyanoquinodimethane (TTF-TCNQ) the best known example—will not be discussed. The interplay between density waves and superconductivity, so-called field-induced spin density waves, and other topics, though very interesting in their own right, will also not be covered by these notes.

I am grateful to several colleagues, in particular to Stuart Brown, Steven Kivelson, George Kriza, Kazumi Maki, Attila Virosztek, and Wolfgang Wonneberger who read and commented upon the various chapters. My students, Steve Donovan, Yong Kim, and Andrew Schwartz were kind enough to take the time and correct many of the mistakes in the early versions of these notes. The first draft was typed by Renée Wellin and the final version by Stella Lozano. The figures were drawn by Jackie Payne. Their help is highly appreciated.

And of course my thanks to Dani, Dora, and Maria—for just being around.

Los Angeles, 1994

Density Waves in Solids

1 The One-Dimensional Electron Gas

Deine Zauber binden Wieder
Was die Mode streng geteilt;

Fashion's laws, indeed may sever,
But thy magic joins again;
—Friedrich Schiller *Hymn of Joy*

Most of the information in subsequent chapters is based on observations made on materials which have a highly anisotropic crystal and electronic structures. These types of materials are usually called "quasi-one-dimensional" or "low-dimensional". The notion refers both to the crystal and to the electronic structure, but it also indicates that concepts characteristic of phenomena which occur in one dimension may often apply.

The reduction of phase space from three dimensions (3D) to one dimension (1D) has several important consequences. Both interaction effects and random potentials have a more profound effect in one than in higher dimensions and fluctuations are also more important. Also, because of the simple Fermi surface in one dimension, the interaction between electrons can be expressed in terms of two coupling constants, one for $q = 0$ and one for $q = 2k_F$; leading to simple phase diagrams for the occurrence of the various broken symmetry ground states which arise as a consequence of these interactions.

1.1 The Response Function of the One-Dimensional Electron Gas

The Fermi surface of a one-dimensional electron gas is simple: it consists of two points, one at $+k_F$ and one at $-k_F$, for an extremely anisotropic metal, two sheets, a distance of $2k_F$ apart. The dispersion relation for a 1D free electron gas is given by $\epsilon(k) = \hbar^2 k^2/2m$, and the Fermi energy by

$$\epsilon_F = \frac{\hbar^2}{2m}\left(\frac{N_0 \pi}{2L}\right)^2 = \frac{\hbar^2 k_F^2}{2m_e} \tag{1.1}$$

where N_0 is the total number of electrons, L is the length of the 1D chain, and m_e is the free electron mass.

The Fermi wavevector is

$$k_F = \frac{N_0 \pi}{2L} = N_e \pi \tag{1.2}$$

where N_e is the number of electrons per unit length and per spin direction. The density of states for one spin direction is

$$n(\epsilon) = \frac{L}{\pi\hbar}\left(\frac{m_e}{2\epsilon}\right)^{1/2} = \frac{L}{\pi\hbar v} \tag{1.3}$$

where the velocity v is given by the relation $m_e v = \hbar k$.

The particular topology of the Fermi surface leads to a response to an external perturbation which is dramatically different from that obtained in higher dimensions. The response of an electron gas to a time independent potential

$$\phi(\vec{r}) = \int_q \phi(\vec{q})\, e^{i\vec{q}\cdot\vec{r}}\, d\vec{q} \tag{1.4}$$

is usually treated within the framework of linear response theory (see, for example, Kittel, 1963). The rearrangement of the charge density, expressed in terms of an induced charge

$$\rho^{\text{ind}}(\vec{r}) = \int_q \rho^{\text{ind}}(\vec{q})\, e^{i\vec{q}\cdot\vec{r}}\, d\vec{q} \tag{1.5}$$

is related to $\phi(\vec{r})$ through

$$\rho^{\text{ind}}(\vec{q}) = \chi(\vec{q})\phi(\vec{q}) \tag{1.6}$$

where $\chi(\vec{q})$, the so-called Lindhard response function, is given in d dimensions by

$$\chi(\vec{q}) = \int \frac{d\vec{k}}{(2\pi)^d} \frac{f_k - f_{k+q}}{\epsilon_k - \epsilon_{k+q}} \tag{1.7}$$

where $f_k = f(\epsilon_k)$ is the Fermi function. For a three-dimensiona

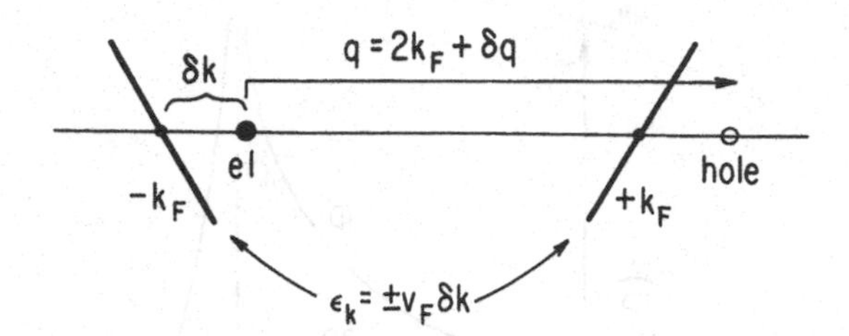

Figure 1.1. The dispersion relation for a free electron gas. The linear dispersion $\epsilon - \epsilon_F = \pm v_F(k - k_F)$ is used to evaluate the response function, Eq. (1.10).

spherical Fermi surface a straightforward calculation gives

$$\chi(\vec{q}) = -e^2 n(\epsilon_F)\left[1 + \frac{1-x^2}{2x}\, ln\left|\frac{1+x}{1-x}\right|\right] \tag{1.8}$$

where $n(\epsilon_F)$ is the density of states at the Fermi level per spin direction, and $x = q/2k_F \cdot \chi(\vec{q})$, as given by Eq. (1.8), decreases with increasing q and the derivative has a logarithmic singularity at $q = 2k_F$.

The situation is different for a one-dimensional electron gas. For wavevectors near $2k_F$, $\chi(q)$ can be evaluated by assuming a linear dispersion relation around the Fermi energy ϵ_F, as shown in Fig. 1.1,

$$\epsilon_k - \epsilon_F = \hbar v_F(k - k_F). \tag{1.9}$$

The integral in Eq. (1.7) can readily be evaluated near $2k_F$ leading to

$$\chi(q) = \frac{-e^2}{\pi\hbar v_F}\, ln\left|\frac{q+2k_F}{q-2k_F}\right| = -e^2 n(\epsilon_F)\, ln\left|\frac{q+2k_F}{q-2k_F}\right|. \tag{1.10}$$

In contrast to a 3D electron gas, the response function in one dimension diverges at $q = 2k_F$. For small q values, $\chi(q)$ is given by the Thomas-Fermi approximation, $\chi(q) = -e^2 n(\epsilon_F)$. The response function, evaluated for all q values, is displayed in Fig. 1.2, where for completeness $\chi(q)$ is also shown for a two- and a three-dimensional electron gas. The fact that $\chi(q)$ diverges for $q = 2k_F$ in the one-dimensional case has several important consequences. Equation (1.6) implies that an external perturbation leads to a divergent charge redistribution; this suggests, through

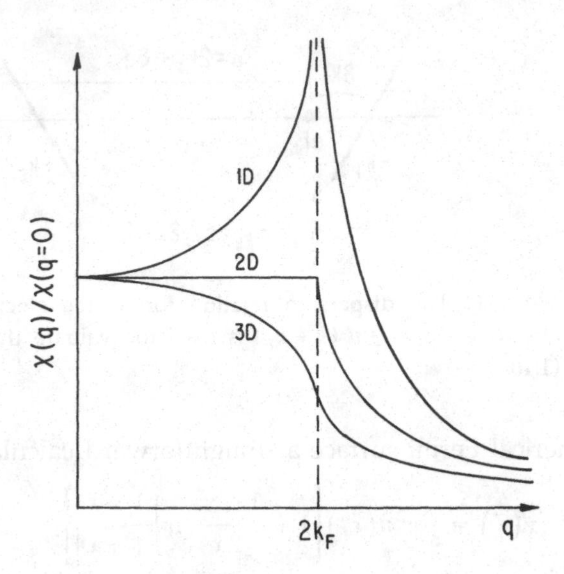

Figure 1.2. Wavevector dependent Lindhard response function for a one-, two-, and three-dimensional free electron gas at zero temperature.

self-consistency, that at $T = 0$ the electron gas itself is unstable with respect to the formation of a periodically varying electron charge or electron spin density. The period is related to k_F by

$$\lambda_0 = \frac{\pi}{k_F}. \tag{1.11}$$

The divergence of the response function at $q = 2k_F$ is due to the particular topology of the Fermi surface, sometimes called perfect nesting. Looking at Eq. (1.7), the most significant contributions to the integral come from pairs of states — one full, one empty — which differ by $q = 2k_F$ and have the same energy, thus giving a divergent contribution to $\chi(q)$. However, in higher dimensions the number of such states is significantly reduced, as shown in Fig. 1.3 leading to the removal of the singularity at $q = 2k_F$. The quasi-one-dimensional character of the Fermi surface can be modeled by including a dispersion in the direction perpendicular to the direction along which the response function was evaluated. The dispersion relation

$$\epsilon(k) = \epsilon_0 + 2t_a \cos k_x a + 2t_b \cos k_y b \tag{1.12}$$

where a and b are the lattice constants in the x and y directions

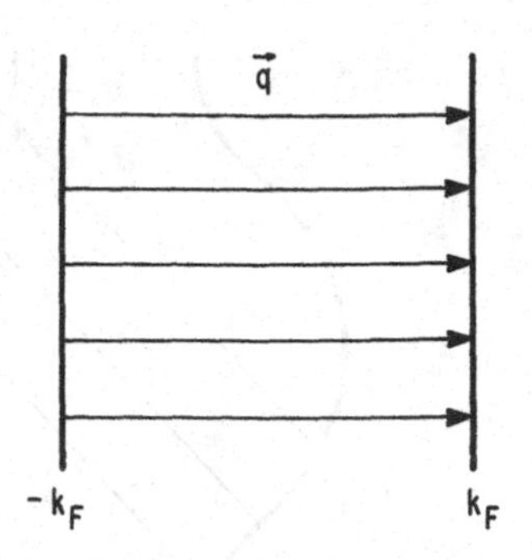

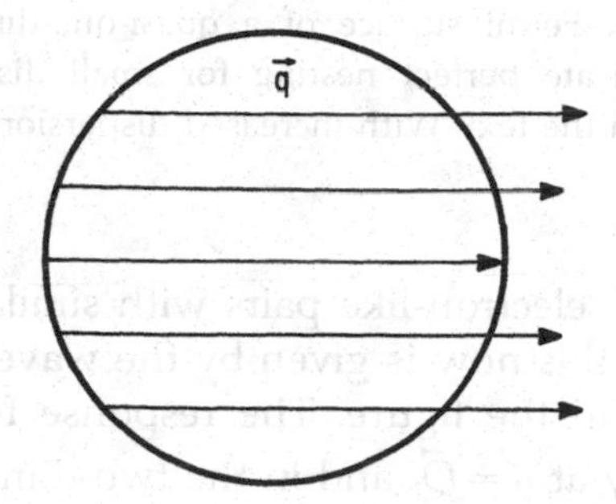

Figure 1.3. Fermi surface topology for a 1D and 2D free electron gas. The arrows indicate pairs of states, one full and one empty, differing by the wavevector $q = 2k_F$.

respectively, leads to a two-dimensional band structure. For $t_a \gg t_b$, and again using a linear dispersion in the x direction this dispersion relation reduces close to the Fermi energy to

$$\epsilon(k) = \epsilon_0 + v_F \delta k - 2t_b \cos k_y b \tag{1.13}$$

with $\delta k = k - k_F$. The Fermi surface is determined by the condition

$$k_x = k_F + \frac{2t_b}{v_F} \cos k_y b + O\left(t_b^2 \cos^2 k_y b\right) + \cdots \tag{1.14}$$

which leads to first the order in t_b (neglecting the third term in Eq. (1.14)), to a sinusoidal Fermi surface in the $k_x - k_y$ plane, as shown in Fig. 1.4. As for one dimension, we recover a large

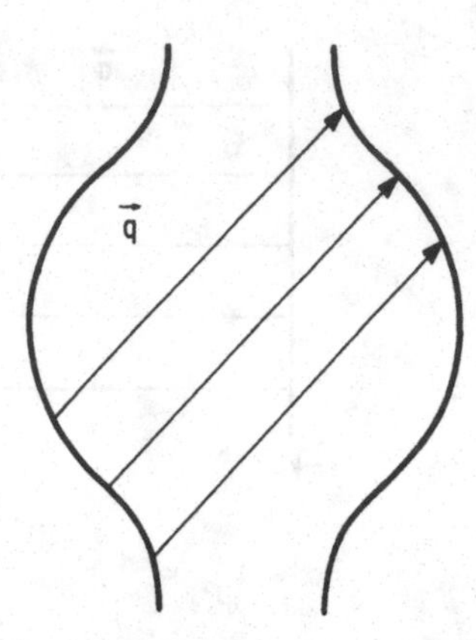

Figure 1.4. Fermi surface of a quasi-one-dimensional electron gas. The arrows indicate perfect nesting for small dispersion in one direction as discussed in the text. With increased dispersion, perfect nesting is no longer possible.

number of electron-like pairs with similar energies; and the condition for this now is given by the wavevector $\vec{Q}=(2k_F,\pi/b)$, as indicated in the figure. The response function $\chi(\vec{q})$ develops a singularity at $\vec{q}=\vec{Q}$, and in the two-dimensional phase space this corresponds to a periodic modulation, with a wavevector $q_\parallel=2k_F$ in the x direction. In the y direction $q_\perp=\pi/b$, and this corresponds to a situation where the modulation on the neighboring chains, separated by b, is out of phase.

Perfect nesting, as shown in Fig. 1.4, is obtained only in the limit when $t_b/t_a \to 0$, and is expected to be appropriate for materials with a substantial anisotropy of the single particle bandwidth. With increasing t_b/t_a the last term in Eq. (1.14) becomes progressively more important and the nesting condition applies for a smaller number of electron-hole pairs. This leads to the gradual removal of the singularity of the response function at $q=2k_F$.

At finite temperatures the numerator in Eq. (1.7) is given by

$$\frac{1}{\exp(-\epsilon_k/k_BT)+1}-\frac{1}{\exp(\epsilon_k/k_BT)+1}=\tanh\frac{\epsilon_k}{2k_BT} \tag{1.15}$$

where ϵ_k is now measured from the Fermi energy ϵ_F. Then

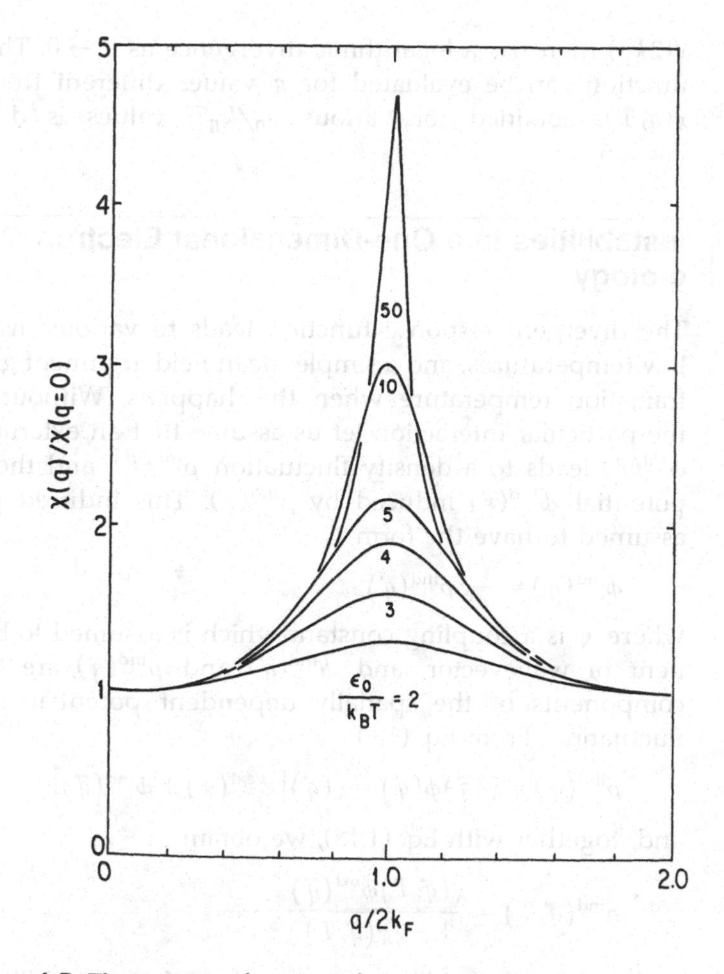

Figure 1.5. The response function of a one-dimensional free electron gas at various temperatures (after Heeger, 1979).

Eq. (1.7) becomes

$$\chi(q = 2k_F, T) = -e^2 n(\epsilon_F) \int_0^{\epsilon_0/2k_B T} \frac{\tanh x}{x}\, dx. \qquad (1.16)$$

Here ϵ_0 is an arbitrarily chosen cutoff energy which is usually taken to be equal to the Fermi energy ϵ_F. The integral can be readily evaluated giving

$$\chi(2k_F, T) = -e^2 n(\epsilon_F) \ln \frac{1.14\epsilon_0}{k_B T}. \qquad (1.17)$$

$\chi(2k_F)$ then has a logarithmic divergence as $T \to 0$. The response function can be evaluated for q values different from $2k_F$ and $\chi(q,T)$, obtained for various ϵ_0/k_BT values is displayed in Fig. 1.5.

1.2 Instabilities in a One-Dimensional Electron Gas: g-ology

The divergent response function leads to various instabilities at low temperatures, and a simple mean field argument gives a finite transition temperature when this happens. Without specifying the particular interaction let us assume that an external potential $\phi^{\text{ext}}(\vec{r})$ leads to a density fluctuation $\rho^{\text{ind}}(\vec{r})$, and therefore to a potential $\phi^{\text{ind}}(\vec{r})$ induced by $\rho^{\text{ind}}(\vec{r})$. This induced potential is assumed to have the form

$$\phi^{\text{ind}}(\vec{q}) = -g\rho^{\text{ind}}(\vec{q}), \tag{1.18}$$

where g is a coupling constant which is assumed to be independent of wavevector, and $\phi^{\text{ind}}(\vec{q})$ and $\rho^{\text{ind}}(\vec{q})$ are the Fourier components of the spatially dependent potential and density fluctuation. From Eq. (1.6)

$$\rho^{\text{ind}}(\vec{q}) = \chi(\vec{q})\phi(\vec{q}) = \chi(\vec{q})\left[\phi^{\text{ext}}(\vec{q}) + \phi^{\text{ind}}(\vec{q})\right] \tag{1.19}$$

and, together with Eq. (1.18), we obtain

$$\rho^{\text{ind}}(\vec{q},T) = \frac{\chi(\vec{q},T)\phi^{\text{ext}}(\vec{q})}{1 + g\chi(\vec{q},T)}. \tag{1.20}$$

The above equation is suggestive of an instability for $g < 0$, which occurs when

$$1 + g\chi(\vec{q},T) = 0. \tag{1.21}$$

Inserting the expression for $\chi(\vec{q},T)$ from Eq. (1.17) into this expression, we obtain

$$1 + gn(\epsilon_F)\ln\frac{1.14\epsilon_0}{k_BT} = 0 \tag{1.22}$$

and the mean field transition temperature is given by

$$k_BT_{\text{MF}} = 1.14\epsilon_0 \exp\left(\frac{-1}{gn(\epsilon_F)}\right). \tag{1.23}$$

As will be discussed later, the density fluctuation $\rho^{\text{ind}}(\vec{q})$, reflects the formation of electron-electron or electron-hole pairs, with the ground state at $T = 0$ being a coherent superposition of the various pair states. The nature of the ground state depends on the electron-electron and electron-phonon interactions, which can be described by a q-dependent interaction potential $V(\vec{q})$.

The situation is particularly simple for a one-dimensional metal where the Fermi surface is two points at $\pm k_F$. (Solyom, 1979 and references cited therein; Emery, 1979). Electrons and holes on the right or left side are denoted by e_+ and e_- and by h_+ and h_-. With the spin degrees of freedom also included, four possibilities for pair formation may occur:

$e_+, \sigma; e_-, -\sigma$	pairs with total momentum with total spin	$q = 0$ $S = 0$
$e_+, \sigma; e_-, \sigma$	pairs with	$q = 0$ $S = 1$
$e_+, \sigma; h_-, \sigma$	pairs with	$q = 2k_F$ $S = 0$
$e_+, \sigma; h_-, -\sigma$	pairs with	$q = 2k_F$ $S = 1$

The development of these states can be discussed using simple models with a mean field solution giving a finite transition temperature. However, as will be discussed later, the mean field solution is not appropriate in one dimension. Solutions of the models which go beyond the mean field treatment lie outside the scope of these notes (see, for example, Solyom, 1979). Consequently, only the main results will be stated here.

The first two of these states develop in response to interactions for which $q = 0$; this is called the particle-particle or Cooper channel. The resulting ground states are the well known (singlet or triplet) superconducting states of metals. The last two states, with a finite total momentum for the pairs, develop as a consequence of the divergence of the fluctuations at $q = 2k_F$; this is the particle-hole channel, usually called the Peierls channel. For these states we find a periodic variation of the charge density or spin density, and consequently, they are called the charge density wave and spin density wave ground states. The period $\lambda_0 = \pi/k_F$ associated with the spatial variation of the charge or spin density

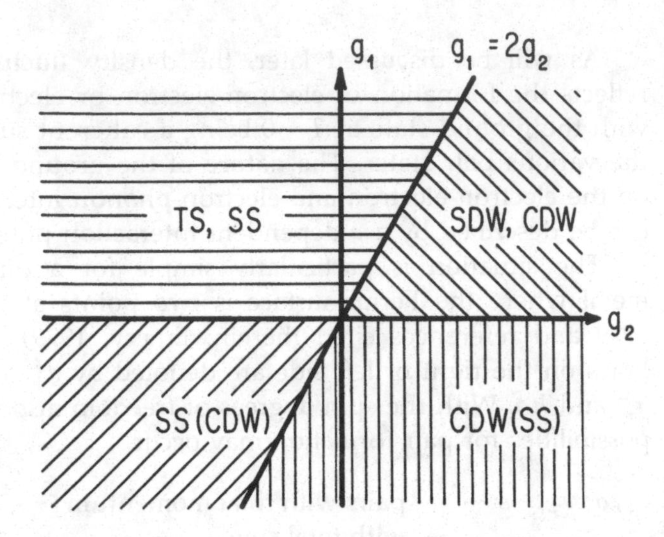

Figure 1.6. The phase diagram of the 1D electron gas in the second order scaling approximation showing the most divergent type of fluctuations, for the coupling constants g_1, g_2. For the definition of the coupling constants and of the various ground states see the text and Solyom, 1979, and references cited therein.

also leads to a gap in the single particle excitation spectrum at the Fermi level. For an arbitrary band filling the period is incommensurate with the underlying lattice.

In 1D the diagrammatic expansion of the Cooper channel contains contributions from the density fluctuations at $q = 2k_F$, and the density fluctuations also contain contributions from the $q = 0$ fluctuations (Solyom, 1979). Consequently, both the Cooper and Peierls response functions diverge at the same rate at low temperatures. Which of these states occurs depends on the detailed nature of electron-phonon and electron-electron interactions, and in one dimension the interaction can be represented as the combination of two coupling constants, g_1 and g_2; which represent the interactions with momentum transfer of $\pm 2k_F$ and zero, respectively. The occurrence of these states in the g_1–g_2 phase space is shown in Fig. 1.6 and this so-called g-ology picture has been discussed in several reviews. Obviously a strictly one-dimensional situation cannot occur in real materials even in compounds with rather highly anisotropic band structure, since interchain interactions are also of importance. Systems which are composed of coupled chains have a phase diagram somewhat different from that shown in Fig. 1.6; but the four different ground states with features rather similar to those calculated for a purely one-dimensional band structure are still preserved for weak interchain coupling.

The various broken symmetry ground states have several common characteristics which are well known for the extensively studied superconducting state (see, for example, Schrieffer, 1963; Tinkham, 1975). For all condensates, the order parameter is complex, and can be written as

$$\Delta = |\Delta| e^{i\phi}. \tag{1.24}$$

For the superconducting ground states, gauge symmetry is broken; the phase is invariant under a gauge transformation. In contrast, for the density wave ground states, the translational symmetry is broken. While the time and spatial derivatives play an important role in the dynamics of the collective modes, the above difference between the superconducting and density wave ground states leads naturally to different collective excitations and also to differences in the coupling of the collective modes to applied electromagnetic fields. For the density wave ground states these collective excitations are called phasons and amplitudons, referring to fluctuations of the phase and amplitude of the condensate. For charge density waves, both occur below the single particle gap, as will be discussed in Chapter 6. In addition, in the spin density wave ground state (as in the triplet superconducting state), the spin rotational symmetry is also broken, leading to additional collective excitations, similar to those of conventional antiferromagnets. The features of the various broken symmetry states are summarized in Table 1.1.

The amplitude $|\Delta|$ is related to (or can be defined as) the single particle gap. In the case of electron-electron pairing this is the well known superconducting gap. In case of the density wave states, the gap occurs at $\pm k_F$. In the superconducting state, the collective mode leads to a supercurrent in response to *dc* fields. In the case of density waves, however, the collective mode does not contribute to the *dc* conduction (this is due to the interaction with impurities and lattice imperfections), and the appearance of a gap in the single particle excitation spectrum at the Fermi level

Table 1.1. Various broken symmetry ground states of one-dimensional metals. The Anderson-Higgs mechanism removes the low lying exitations in a singlet superconductor.

	Paring	*Total Spin*	*Total Momentum*	*Broken Symmetry*	*Low Lying Collective Excitations*
singlet superconductor	el-el	$S=0$	$q=0$	gauge	(Anderson-Higgs)
triplet superconductor	el-el	$S=1$	$q=0$	gauge	low lying magnetic excitations?
charge density wave	el-hole	$S=0$	$q=2k_F$	translational	phasons amplitudons
spin density wave	el-hole	$S=1$	$q=2k_F$	translational	phasons magnons

causes the material to become a semiconductor below the transition temperature.

The appearance of the single particle gap Δ also leads to a finite coherence length ξ_0, which corresponds to the spatial dimension of the electron-electron or electron-hole pairs. Crudely speaking, the pair wavefunctions are the superpositions of one electron state within the energy region around the Fermi level, and the corresponding spread of momenta is approximately

$$|\Delta| = \delta\left(\frac{p^2}{2m}\right) \simeq \hbar v_F \delta_p \tag{1.25}$$

where v_F is the Fermi velocity. This corresponds to a spatial range of $\xi_0 \simeq (\delta_p)^{-1} = \hbar v_F/\Delta$. The correct expression, the BCS coherence length, is given at zero temperature by (Schrieffer, 1964; Tinkham, 1975)

$$\xi_0 = \frac{\hbar v_F}{\pi|\Delta|}. \tag{1.26}$$

The temperature dependence of the order parameter and condensate density have the well known BCS form for all cases; and the gap is related to the mean field transition temperature, T_{MF}, through the well known BCS relation $2|\Delta|(T=0) = 3.52 k_B T_{MF}$ in the weak coupling limit. The condition when this limit applies is somewhat different for different ground states, as will be discussed later.

1.3 Correlations and Fluctuations

Because of the reduction of phase space, one-dimensional systems are unstable against fluctuations. These fluctuations lead to the absence of long range order at any finite temperature, and for $T \neq 0$ only short range correlations develop. The correlation length for models which involve the response of the electron gas to electron-phonon or electron-electron interactions, is related to the correlation length which characterizes the density fluctuations of the one-dimensional electron gas. These density fluctuations are described by the correlation function

$$\langle C^{\dagger}(\vec{r}), C(0)\rangle = \int \frac{d\vec{k}}{2\pi} f(\epsilon_k)\, e^{ik\cdot r} \tag{1.27}$$

where $C^{\dagger}$ and C are the creation and destruction operators of the electron density. The spatial dependence is approximately given by

$$\langle C^{\dagger}(\vec{r}), C(0)\rangle \simeq \exp\left[\frac{|\vec{r}-\vec{r}'|}{\xi} + i\vec{q}\cdot\vec{r}\right] \tag{1.27a}$$

where ξ is the correlation length associated with the density fluctuations. For a one-dimensional metal, with a linear dispersion relation $\epsilon_k - \epsilon_F = \hbar v_F(k - k_F)$ as shown in Fig. 1.1,

$$\begin{aligned}\langle C^{\dagger}(\vec{r}), C(0)\rangle &= \int \frac{dk}{2\pi} \frac{e^{ikx'}}{e^{\beta v_F |k-k_F|} + 1} \\ &= \frac{ie^{ik_F x}}{-\beta v_F} \sum_{j=0}^{\infty} e^{-\pi(2j+1)x/\beta v_F} \\ &= \frac{ie^{ik_F x}}{-\beta v_F} \frac{e^{-\pi x/\beta v_F}}{e^{-2\pi x/\beta v_F} + 1}\end{aligned} \tag{1.28}$$

where $\beta = (k_B T)^{-1}$, and using Eq. (1.27a) the correlation length

$$\xi = \frac{\hbar v_F}{\pi k_B T} \tag{1.29}$$

diverges as the temperature approaches $T = 0$.

As will be discussed later, at low temperatures the collective excitations of the density wave states are described by the Hamiltonian

$$\mathscr{H} = A\int (\nabla\phi)^2\, d\vec{r} = \int d^d\vec{k}\, |\phi_k|^2 k^2 \tag{1.30}$$

in dimension d where ϕ is the phase of the condensate and A is the elastic constant associated with the long wavelength deformations of the condensate. The thermal expectation value of the component $|\phi_k|^2$ is

$$\langle |\phi_k|^2 \rangle = \frac{\int d\phi_k \exp\left(-\dfrac{k^2|\phi_k|^2}{k_B T}\right)|\phi_k|^2}{\int d\phi_k \exp\left(-\dfrac{k^2|\phi_k|^2}{k_B T}\right)} \tag{1.31}$$

which, after factorization becomes approximately

$$\langle |\phi_k|^2 \rangle \simeq \frac{k_B T}{k^2}. \tag{1.32}$$

Let us look at the correlation function which describes the spatial fluctuations of the phase variable. In the limit when the amplitude of the order parameter is close to its $T=0$ value, the correlation function looks like

$$\langle \Delta^*(\vec{r}), \Delta(0) \rangle = |\Delta_0|^2 \exp - \frac{1}{2}\left\langle [\phi(\vec{r}) - \phi(0)]^2 \right\rangle; \tag{1.33}$$

the term in the exponent,

$$\left\langle [\phi(\vec{r}) - \phi(o)]^2 \right\rangle = \int_k d^d\vec{k} \langle |\phi_k|^2 \rangle (e^{i\vec{k}\cdot\vec{r}} - 1)^2 \tag{1.34}$$

$$= k_B T \int \frac{d^d\vec{k}}{k^2} (e^{i\vec{k}\cdot\vec{r}} - 1)^2.$$

The integral diverges for $d \leq 2$, and therefore there is no long range order at any finite temperature.

The argument advanced here is essentially the same as the argument for the absence of long range order for a Heisenberg antiferromagnet, and is due to the existence of low lying, gapless collective modes. If these excitations develop a gap (and this would occur for a commensurate density wave, as will be discussed in Chapter 7) long range order is restored. Arguments, analogous to those applied for the Ising model (Ziman, 1964) apply for this latter situation and there is no long range order for $d \leq 1$ at finite temperatures.

Materials

Néhány Anyag
más-más tulajdonságokkal felruházva
Substances, each
blessed with different attributes

—Imre Madach *The Tragedy of Man*

2

A large number of organic and inorganic solids have crystal structures in which the fundamental structural units form linear chains. While most of these materials are insulators or semiconductors; several groups have partially filled electron bands, and consequently display metallic behavior at high temperatures.

The greatly different overlap of the electronic wave functions in the various crystallographic directions leads to strongly anisotropic, so-called quasi-one-dimensional electron bands discussed in the previous chapter. This is the prerequisite for the development of the instabilities at $q = 2k_F$. Among the different materials which display a variety of phase transitions, only the simplest examples of density wave formation will be discussed, where the charge or spin density wave fluctuations develop along identical chains and, subsequently, the interaction between the chains leads to a three-dimensional ordered ground state. The partially filled electron bands, where the number of electrons per site is not a simple fractional number (eg., 1/2, 1/3, etc.), lead to density wave instabilities where the period $\lambda_0 = \pi / k_F$ of the density waves is incommensurate with the underlying lattice (for which the fundamental period is the lattice constant a). Materials with two-dimensional band structures (and consequently with 2D density wave ground states) will not be discussed. Materials which are composed of two or more different metallic chains (consequently displaying a variety of subtle instabilities) also lie outside the scope of this discussion.

While probably a coincidence, among the various compounds with a single conducting chain, *inorganic* linear chain compounds have been found to be examples of charge density wave condensates; while several groups of *organic* materials have been observed to develop a spin density wave ground state. In both cases a broad variety of experiments have been conducted to explore the normal state properties, with a focus on the parameters which characterize the single particle electron states, on the anisotropy, and on the strength of the electron-electron and electron-phonon interactions.

The crystal structures are in general complex. Several reviews listed in the Appendix discuss the structural properties of the various groups of linear chain materials in detail. Therefore only the basic features of the structural arrangements will be summarized here, with emphasis on the resulting electronic structure of these types of materials.

Because of the strong anisotropy, in many cases the bandwidths in the directions perpendicular to the chains are smaller than, or comparable to, the thermal energy k_BT at room temperature (or at temperatures above the density wave transition). Thus, expressions appropriate for a 1D electron gas can be used to evaluate parameters such as the Fermi energy ϵ_F, the Fermi velocity v_F, and the bandmass m_b, which can be extracted by using various experimentally accessible quantities. For a 1D band structure

$$k_F = \frac{\pi n}{a}; \qquad v_F = \frac{\hbar k_F}{m_b} = \frac{\hbar n \pi}{a m_b} \tag{2.1}$$

where n is the number of electrons per unit cell and per spin direction. (The unit cell in one-dimension is given by the lattice constant a.) Deviations from the free electron behavior are usually included by assuming a bandmass m_b which is different from the free electron mass m_e. In the above equation n refers to the number of electrons for the 1D electron band, and is given by

$n = N_e a$, the total number of electrons per unit volume is given by $n_0 = N_e n_\perp$, with $n_\perp$ being the number of chains per unit area. Then, the density of states $n(\epsilon_F)$ for each spin direction is related to ϵ_F by

$$n(\epsilon_F) = \frac{N_e}{4\epsilon_F} = \frac{n}{4a\epsilon_F}. \tag{2.2}$$

In general, n (or N_e, the number of electrons per unit length) can be derived from electron counting arguments, as well as from the measured period $\lambda_0 = \pi/k_F$ of the density wave. If available from experiments, the plasma frequency

$$\omega_p = \left(\frac{4\pi ne^2}{m_b}\right)^{1/2} \tag{2.3}$$

can then be used to evaluate the bandmass. With n and m_b known, Eqs. (2.1) and (2.2) can be used to determine the Fermi energy ϵ_F.

The magnetic susceptibility

$$\chi_p = g^2\mu_B^2 n(\epsilon_F) = g^2\frac{\mu_B^2 N_e}{2\epsilon_F} = g^2\frac{\mu_B^2 n}{2a\epsilon_F} \tag{2.4}$$

where g is the gyromagnetic factor and μ_B the Bohr magneton, can also be used; together with k_F as obtained from the calculated band filling or from the period λ_0 of the measured lattice distortion, to derive the same parameters. Broadly speaking as electron-electron interactions lead to an enhancement of the magnetic susceptibility while ω_p does not reflect these interactions, the comparison of the two sets of parameters may also shed light on the importance of these interactions.

The above equations are based on a nearly free electron approach; a tight binding description is also often used to extract parameters such as the Fermi energy ϵ_F and Fermi velocity v_F from the measured quantities. The two approaches lead to somewhat different values for the Fermi energy, Fermi velocity, and density of states. These differences, however, are not essential, particularly when one is interested only in the gross overall features of the phase transitions, and in the approximate values of the parameters which characterize the various broken symmetry ground states.

The expressions in Eqs. (2.1) and (2.2) which lead to the Fermi energy and Fermi velocity from measured quantities such as ω_p and χ are obviously appropriate only for a strictly 1D electron band, and should consequently be used only for a rather anisotropic bandwidth. This anisotropy can be estimated by using the measured anisotropy of the *dc* electrical conductivity σ_{dc}, and crude arguments (Jerome and Schultz, 1982) lead to

$$\frac{\sigma_{\|}}{\sigma_{\perp}} \simeq \left(\frac{v_{F\|}}{v_{F\perp}}\right)^2 \tag{2.5}$$

where $v_{F\|}$ and $v_{F\perp}$ are the Fermi velocities along and perpendicular to the chain direction. The anisotropy of the optical reflectivity, in particular the anisotropic plasma frequency (if it exists in both directions), has also been used to assess the magnitude of the anisotropy of these materials.

2.1 Inorganic Linear Chain Compounds

A variety of inorganic materials have strongly anisotropic crystal, and consequently strongly anisotropic electronic structures; which often display various transitions from metallic to nonmetallic states. Among the many groups of such materials, three have been explored in detail and these materials are by now well established examples of the charge density wave ground state.

2.1.1 Mixed Valence Platinum Chain Compounds

Platinum chain complexes are composed of a columnar array of units which incorporate a chain of Pt atoms with strongly overlapping *d* orbitals. Although a considerable number of compounds of such structure are known, most of the experiments have been performed on the material $K_2Pt(CN)_4Br_{0.3} \cdot 3.2H_2O$, usually called simply KCP or Krogmann's salt. The schematic crystal structure of this material is shown in Fig. 2.1. It consists of a columnar stacked array of $Pt(CN)_4$ units, with a rather short Pt–Pt separation of 2.894 Å along the chain direction. The distance between the chains in 9.89 Å leading to a strongly anisotropic separation between the Pt atoms in the different directions. The water molecules form a hydrogen-bonded material between the

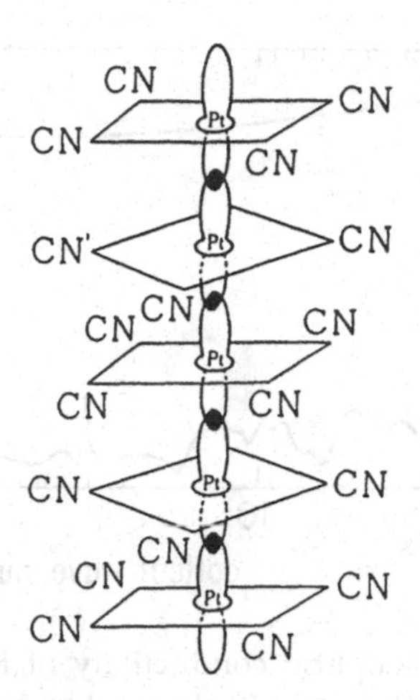

Figure 2.1. The crystal structure of the compound $K_2Pt(CN)_4Br_{0.3} \cdot 3.2H_2O$, called KCP, or Krogmann's salt.

CN^- ligands and the K^+ ions provide a cross-link between the $Pt(CN)_4$ chains.

The $K_2^{2+}[Pt(CN)_4]^{2-}H_2O$ configuration has a full valency Pt^{2+} and consequently this material would be a semiconductor. The Br counterions remove electrons from the $Pt(CN)_4$ unit, and the resulting fractional charge $Pt^{1.7}(CN)_4$ leads to a partially filled electron band, the prerequisite for metallic behavior.

Indeed, the material is highly conducting at room temperature, with a conductivity along the chain direction of $\sigma_{\|} \sim 10^2 \Omega^{-1}$ cm^{-1} (Carneiro, 1988). The conductivity is also highly anisotropic, and the ratio of the conductivities measured along and perpendicular to the chains is $\sigma_{\|}/\sigma_{\perp} \sim 10^5$. This anisotropy is also clearly evident in the optical properties (Bruesh et al., 1975; Geserich, 1988), and the reflectivity measured in the two different directions is displayed in Fig. 2.2. For $E \| c$, the optical conductivity is well described by the usual Drude form

$$\sigma(\omega) = \frac{\sigma_{dc}}{1 - i\omega\tau} = \frac{\omega_p^2 \tau}{2\pi} \frac{1}{(1 - i\omega\tau)} \tag{2.6}$$

with a plasma frequency of $\omega_p = 2.9$ eV and a relaxation time of $\tau = 3 \times 10^{-15}$ sec. The band filling of $1.7/2$ $el = 0.85$, obtained from charge transfer arguments as discussed before, gives a band-

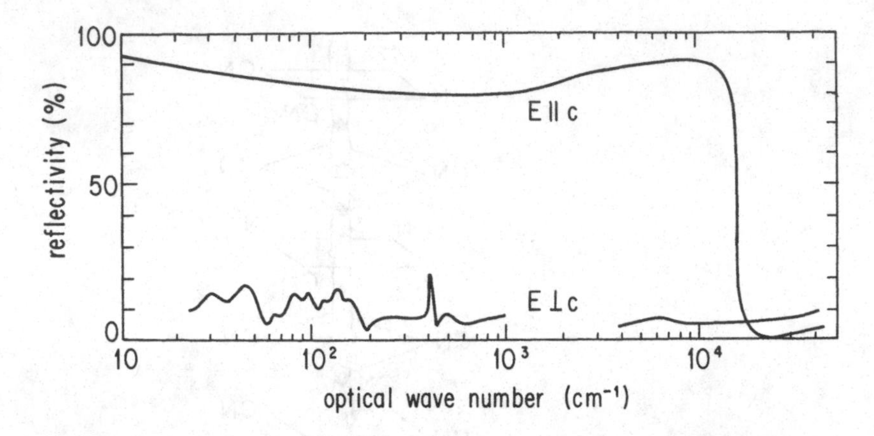

Figure 2.2. The optical conductivity of $K_2Pt(CN)_4Br_{0.3} \cdot 3.2H_2O$ measured both parallel and perpendicular to the chain direction at room temperature (after Geserich, 1988); c refers to the chain direction.

mass of $m_b \simeq 1.0\ m_e$ as expected for strongly overlapping Pt orbitals forming wide conduction bands. This relatively wide electron band along the chain direction also follows from the small Pauli susceptibility measured at room temperature or above (Scott et al., 1979). These magnetic measurements also indicate that electron-electron interactions do not play an important role in these materials.

2.1.2 Transition Metalchalcogenides, MX_3 and $(MX_4)_nY$

Group IV or V transition metals, Nb or Ta, when combined with chalcogen atoms, S or Se, form a variety of linear chain compounds, several of which have partially filled electron bands and thus metallic behavior at high temperatures. (Rouxel and Schlenker, 1989; Meerschaut and Rouxel, 1976).

The basic constituent of the structure is a triangular prism of MX_6 units, with a cross-section close to an isosceles triangle as shown in Fig. 2.3. The transition metal atom (indicated by the solid circle in the figure) is located roughly at the center of the prism. The prisms are stacked on top of each other by sharing the triangular faces along the b-axis, and the chains are staggered with respect to each other by half the height of the unit prism. Therefore, besides the six chalcogen atoms of an MX_6 prism, each transition metal is bonded to two more X atoms from neighbor-

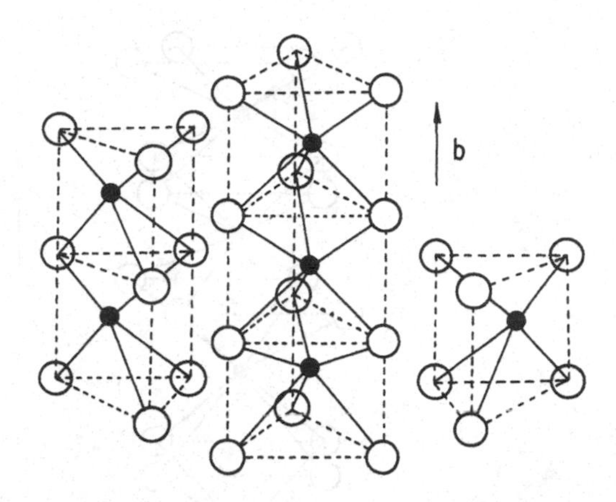

Figure 2.3. The schematic chain structure of the MX_3 compounds. The M atoms are located roughly at the center of trigonal prisms built of atoms X.

ing chains, and its coordination number is eight. Various chains in the crystal structure have slightly different geometrical configurations, and the number of different chains varies from compound to compound. In $NbSe_3$ and TaS_3 three different chains are observed. The latter compound also occurs in two different crystallographic (monoclinic and orthorhombic) modifications.

In these compounds, the transition metal atoms have an electron configuration $4d^3 5s^1$ and some of the chalcogen atoms form covalently bonded pairs. The electron configuration can schematically be written as

$$3MX_3 = 2M^{5+}M^{4+}(X^{2-})_5(X_2^{2-})_2$$

which is suggestive of a quarter filled electron band involving the dorbitals of the transition metal atoms.

The schematic crystal structure of the tetrachalcogenides $(MX_4)_nY$ with $M =$ Ta or Nb, $X =$ S or Se, and $Y =$ I, Br or Cl is displayed in Fig. 2.4. (Gressier et al., 1984). Each transition metal (indicated by the solid circle on the figure) is coordinated to eight Se atoms (open circles) in a slightly distorted rectangular antiprismatic arrangement. Some of the materials undergo structural distortions (unrelated to CDW formation), but in the compounds $(TaSe_4)_2I$ and $(NbSe_4)_2I$ such distortions do not occur and these materials are relatively good metals at high temperatures. Al-

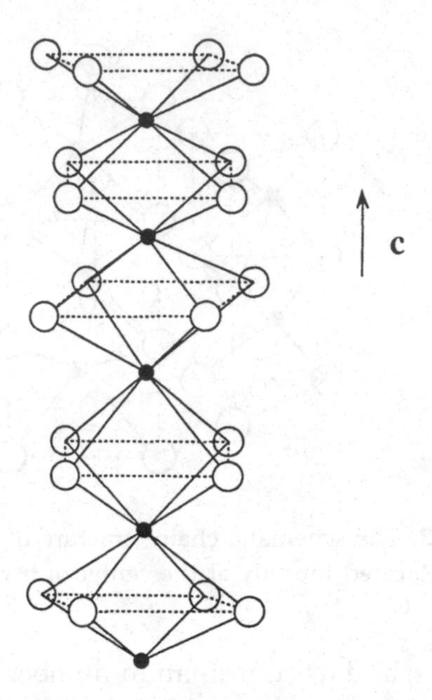

Figure 2.4. The schematic chain structure of the $(MX_4)_2Y$ compounds.

though the former compound may be formulated as $Ta^{4+}Ta^{5+}4[Se_2]^{2-}I$ from charge neutrality arguments, all Ta–Ta distances are found to be identical (3.206 Å) along the chain and the electron band is quarter filled; the same applies to the Nb compound. The Se atoms lie in planes approximately perpendicular to the c-axis (which is the chain direction in this case). These planes (separated by $c/4$) develop in a helicoidal way along the c-axis. The chains of $TaSe_4$ are parallel to the c direction and are well separated by the I atoms. The materials are relatively good metals along the chain direction with room temperature conductivities parallel to the chains, $\sigma_{\|}$, on the order of 10^3–$10^4 \Omega^{-1}$ cm^{-1}. The conductivity perpendicular to the chains is 10 to 10^3 times smaller, suggesting that the anisotropy of the bandwidth is substantial.

2.1.3 Transition Metal Bronzes

The term bronze is applied to a variety of crystalline phases of the transition metal oxides. Examples are the ternary molybdenum oxides of formula $A_{0.3}MoO_3$, where the alkali metal A can be K,

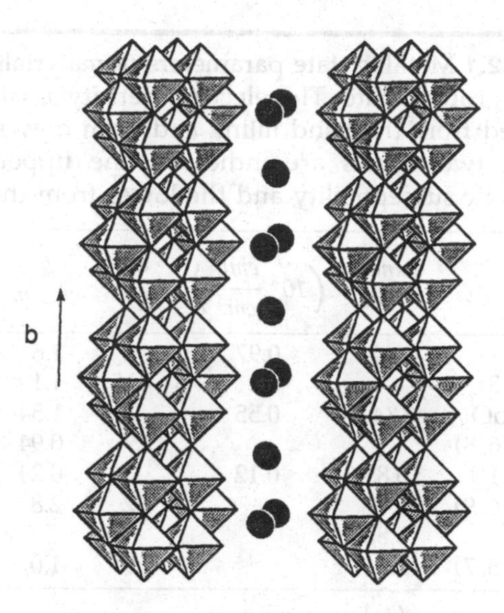

Figure 2.5. The schematic structure of $K_{0.3}MoO_3$.

Rb or T*l*. They are often referred to as blue bronzes because of their deep blue color (Schlenker and Dumas, 1986).

The structure of $A_{0.3}MoO_3$ contains rigid units comprised of clusters of ten distorted MoO_6 octahedra, sharing corners along the monoclinic *b*-axis, as illustrated in Fig. 2.5. This corner sharing provides an easy path for the conduction electrons along the chain direction. The chains of distorted MoO_6 octahedra also share corners along the [102] direction and form infinite slabs separated by the alkaline cations.

The band structure, calculated for a slab of $Mo_{10}O_{30}$ clusters is fairly complicated (Whangboo and Schneemeyer, 1986), but such calculations give two overlapping bands at the Fermi level, both approximately 3/4 filled. This, as will be discussed in Chapter 3, is in agreement with the lattice distortion observed in the charge density wave state of the material. The Fermi surface has a pronounced two-dimensional character in this compound as evidenced by *dc* conductivity and optical studies. At room temperature, the conductivity measured along the chain direction is $\sigma_b = 3 \times 10^2 \Omega^{-1}$ cm^{-1}. In the two perpendicular directions the conductivity is $\sigma_{2a-c} = 10\Omega^{-1}$ cm^{-1} and $\sigma_{2a+c} = 0.5\Omega^{-1}$ cm^{-1}. The crude argument which gives Eq. (2.5) then leads to an

Table 2.1 Metallic state parameters of materials with a charge density wave ground state. The electron density n (not displayed) has been derived from the band filling and from crystal structure parameters. When two values are indicated, the upper is derived from the magnetic susceptibility and the lower from the plasma frequency.

	Band filling	χ $\left(10^6 \frac{emu}{cm^3}\right)$	ω_p (eV)	m_b/m_e	ϵ_F (eV)	v_F (10^7 cm/sec)
$NbSe_3$	1/4	0.97		1.8	0.11	1.4
(ref. 1,5)			1.15	4.1	0.048	0.65
$K_{0.3}MoO_3$	3/4	0.55		1.54	0.24	2.3
(ref. 2,6,8)			2.7	0.94	0.39	3.8
$(TaSe_4)_2I$	0.85	0.12		0.24	0.68	9.9
(ref. 3,6,9)			1.2	2.8	0.059	0.87
KCP	0.85					
(ref. 4,5,7)			2.9	1.0	0.82	5.37

	$n(\epsilon_F)$ (eV^{-1})	ω_{2k_F} (eV)
$NbSe_3$	2.3	
(ref. 1,5)	5.2	
$K_{0.3}MoO_3$	3.1	
(ref. 2,6)	1.9	3×10^{-3} (ref. 8)
$(TaSe_4)_2I$	1.3	
(ref. 3,6)	14.4	
KCP		
(ref. 4,5,7)	1.0	8×10^{-3} (ref. 9)

1. H. P. Geserich, et al. Physica **143B**, 174 (1986).
2. G. Travaglini and P. Wachter, Solid State Comm. **37**, 599 (1981).
3. H. P. Geserich, et al. Physica **143B**, 198 (1986).
4. P. Bruesh, et al. Phys. Rev. **B12**, 219 (1975).
5. J. C. Scott, et al. Phys. Rev. **B10**, 3131 (1979).
6. D. Johnston, et al. Solid State Comm. **53**, 5 (1985).
7. K. Carneiro, et al. Phys. Rev. **B13**, 4758 (1976).
8. J. P. Pouget and P. Comes, In "Charge Density Waves in Solids" Eds. L. P. Gor'kov and G. Grüner, North Holland 1987.
9. S. Sugai, Physica **143B**, 195 (1986). H. Fujishita, et al. Physica **143B**, 201 (1986).

anisotropy of the Fermi velocity of approximately 6 and approximately 25 for the two perpendicular directions. This has also been confirmed by optical studies (Travaglini et al., 1981).

The parameters of the single particle bands which occur along the chain directions are summarized in Table 2.1. In all cases the strong overlap of the wavefunctions leads to wide bands, with bandwidths significantly larger than the energy scale which, as will be discussed in Chapter 3, corresponds to the single particle gaps associated with the formation of the charge density wave ground state. The relatively low Pauli susceptibility also indicates that electron-electron interactions are small in these materials.

Because of the complicated crystal structures, the phonon spectrum of these materials is also fairly complicated. Which of the phonon modes couples to the electronic degrees of freedom is ultimately determined by neutron scattering experiments. The unrenormalized phonon frequency $\omega(2k_F)$ for the wavevector $q = 2k_F$ can be estimated from these experiments, and the various values are also displayed in Table 2.1.

2.2 Organic Linear Chain Compounds

Planar organic molecules often form linear chains with large overlap of the π orbitals along the chain direction. When combined with counterions or molecules, the resulting charge transfer salts may have partially filled bands leading to metallic properties. Some members of three groups of materials, based on the organic molecules M shown in Fig. 2.6 and having the composition MX_2, also develop a spin density wave ground state at low temperatures as established by a wide range of magnetic measurements.

The crystal structure of the material (tetramethyltetraselenafulvalene)$_2$PF$_6$, (TMTSF)$_2$PF$_6$, one member of the so-called Bechgaard salts (Bechgaard et al., 1980), is shown in Fig. 2.7. The material is built up of segregated stacks of TMTSF and PF$_6$ molecules. The structures of the compounds based on the MDT-TTF and DMET molecules have the same basic features. All three molecules shown in Fig. 2.6 are good donors; and when combined with strong acceptors such as PF$_6$ and similar species, a charge transfer occurs from the M stacks to the X stacks. For a full charge transfer the M stacks become $3/4$ filled. There is a

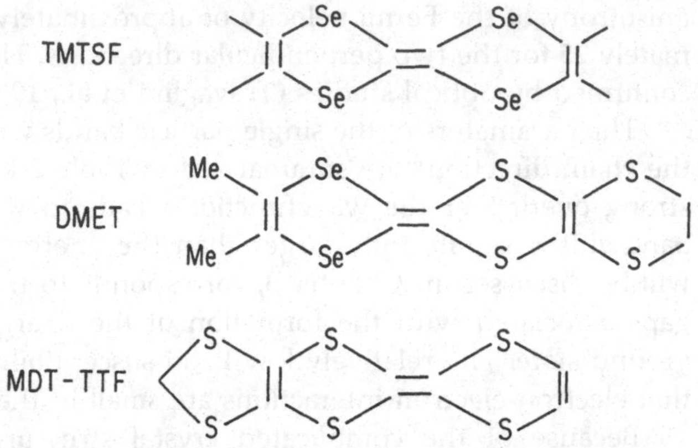

Figure 2.6. Planar organic acceptor molecules which form metallic charge transfer salts, and which undergo a transition to a spin density wave ground state at low temperatures. Me refers to methyl groups and the open ended lines to H.

significant overlap of electronic wavefunctions on the *M* stacks, with practically no overlap along the *X* stacks; consequently, band theory predicts metallic behavior along the chains. The materials which belong to the different groups do indeed show metallic conduction down to low temperatures, as shown in Fig. 2.8 (Bechgaard et al., 1980). The increase of resistivity at low temperatures is due to the removal of the Fermi surface upon the formation of the spin density wave ground state, as will be discussed in Chapter 4. The optical properties of $(TMTSF)_2PF_6$ are that of a Drude metal (Jacobsen et al., 1983), with high reflectivity R along the chain direction ($E\|a$), as shown in Fig. 2.8. In contrast, perpendicular to the chains ($E\|b$) nonmetallic reflectivity is observed at high temperatures. From the analysis of the optical properties the plasma frequencies can be extracted and these parameters, together with electron concentration, are displayed in Table 2.2. The magnetic susceptibility is of the Pauli type (Jerome and Schultz, 1982) and is weakly temperature dependent down to low temperatures where the phase transition occurs. The parameters ω_p and χ can then be used, together with the free electron expressions for these parameters, to extract the parameters which characterize the single particle states, Fermi velocity v_F, Fermi energy ϵ_F, and the density of states $n(\epsilon_F)$.

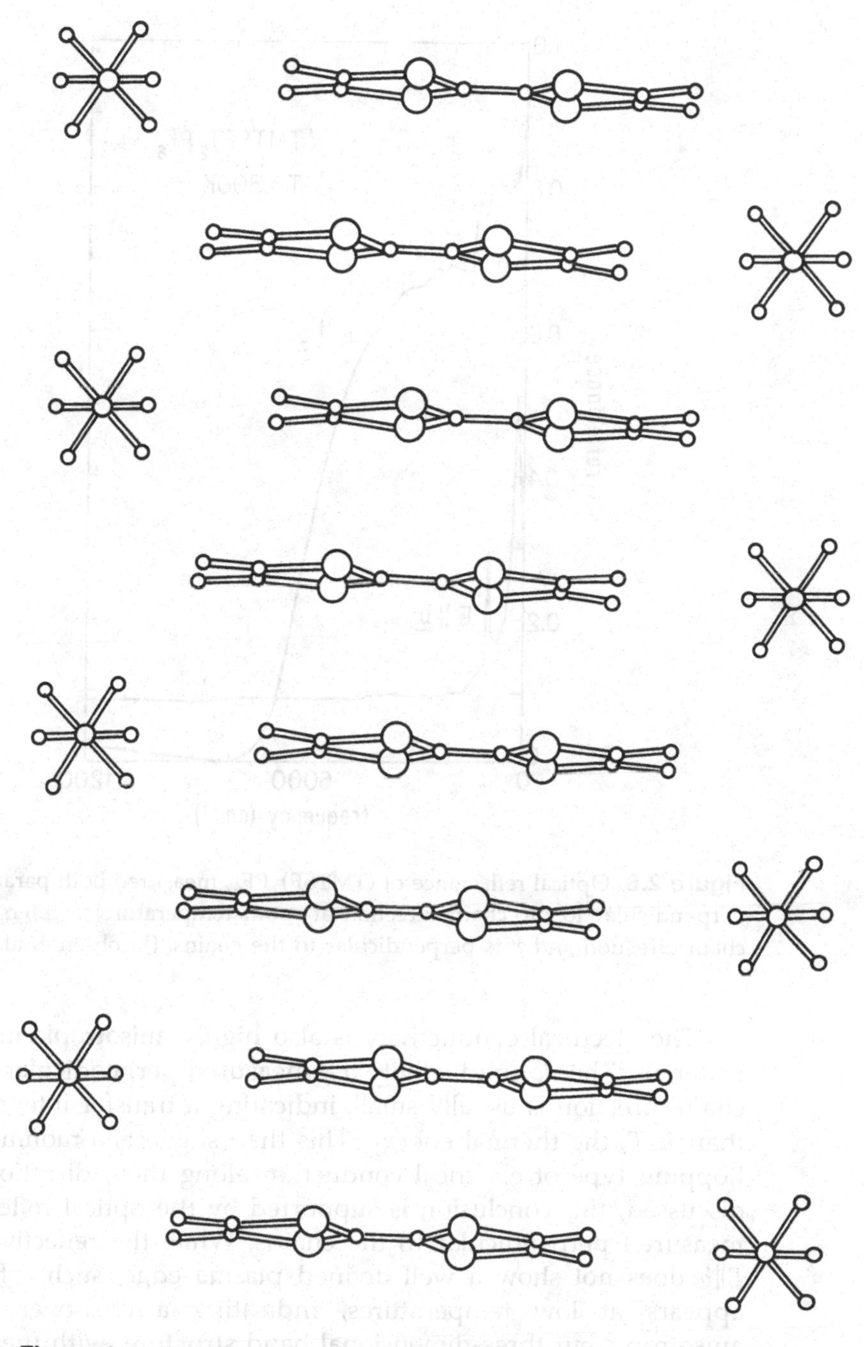

Figure 2.7. The schematic crystal structure of (tetramethyltetraselenafulvalene)$_2$PF$_6$, (TMTSF)$_2$PF$_6$.

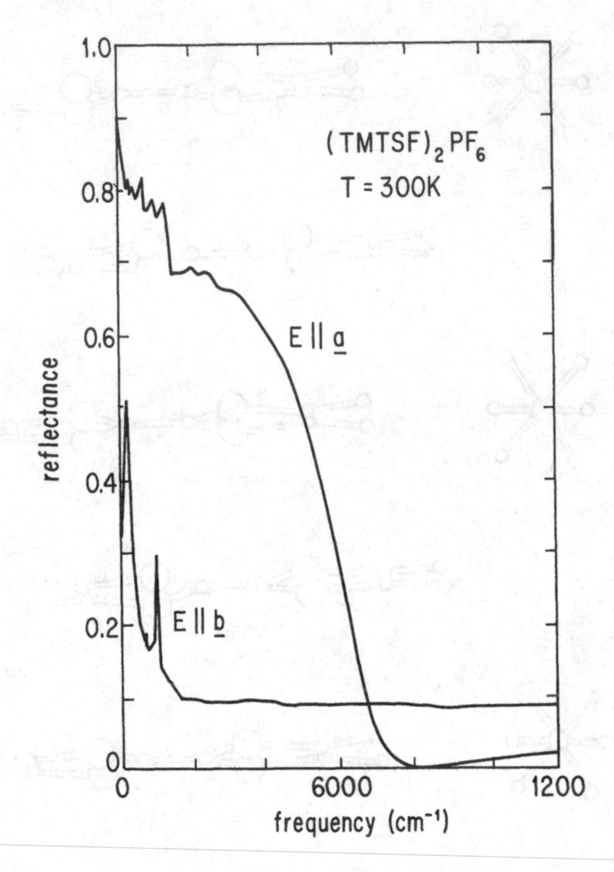

Figure 2.8. Optical reflectance of $(TMTSF)_2PF_6$, measured both parallel and perpendicular to the chain direction at room temperature; *a* refers to the chain direction and *b* is perpendicular to the chains. (Jacobsen et al., 1982).

The electrical conductivity is also highly anisotropic in these materials. The *dc* conductivity σ_{dc} measured perpendicular to the chain direction is usually small, indicating a transfer integral less than $k_B T$, the thermal energy. This then suggests a monmetallic, hopping type of electrical conduction along these directions. As discussed, this conclusion is supported by the optical reflectivity measured perpendicular to the chains. While the reflectivity for $E \| b$ does not show a well defined plasma edge, such a feature appears at low temperatures, indicating a crossover to an anisotropic but three-dimensional band structure, with the bandwidth larger than $k_B T$ in both directions. From these studies,

Table 2.2 Parameters of the metallic state of two materials with a spin density wave ground state. The electron density has been evaluated from the band filling and crystal structure parameters.

	Band filling	χ $\left(10^{-6}\frac{emu}{cm^3}\right)$	ω_p (eV)	m_b/m_e	ϵ_F (eV)
$(TMTSF)_2PF_6$	1/4	0.58 (ref. 1)	2.9 (ref. 2)	1.3	6.1×10^{-2}
$(DMET)_2Au(CN)_2$	1/4	0.51 (ref. 3)		2.4	6.6×10^{-2}

	v_F $\left(10^7\frac{cm}{sec}\right)$	$n(\epsilon_F)$ (eV^{-1})
$(TMTSF)_2PF_6$	0.86	0.25
$(DMET)_2Au(CN)_2$	1.0	0.26

1. K. Mortensen, et al., Phys. Rev. **B25**, 3319 (1982), D. Jerome and H. Schultz, Adv. Phys. **32**, 299 (1982).
2. A Jacobsen, et al., Phys. Rev. **B28**, 7019 (1983).
3. K. Kanoda, et al., Phys. Rev. **B38**, 39 (1988).

bandwidths of 25 meV and 1 meV are inferred for the two perpendicular directions, in contrast to the bandwidth along the TMTSF stacks which is on the order of 0.25 eV (Jerome and Schultz, 1982).

Members of other groups of organic linear chain compounds have also been found to develop a spin density wave ground state. The organic salts, $(MDT\text{–}TTF)_2X$, where MDT–TTF stands for methilendithio-tetrathiafulvanele (shown in Fig. 2.6) have a metallic character for various counterions, and one member of the group, $(MDT\text{–}TTF)_2Au(CN)_2$, has a spin density wave transition at $T = 20K$ (Nakamura et al., 1990). This material, along with other members of the group, has a linear chain structure formed by MDT–TTF stacks with strongly overlapping orbitals in the stack direction. This then leads to a strongly anisotropic single particle band.

The molecule dimethylethylenedithio-diselenadithia-fulvalene, DMET also forms various charge transfer salts with different counterions and the composition is $(DMET)_2X$ (Kikuchi et al., 1987; Kanoda et al., 1988). Some members of this family are semiconductors while others remain metals to low temperatures where they undergo transitions to various broken symmetry states. In the compound $(DMET)_2Au(CN)_2$ the ground state is a spin density wave which develops at $T = 20K$ where a metal-to-insulator transition occurs with decreasing temperatures.

Optical studies have not been performed on these latter two groups of compounds, and magnetic susceptibility has only been measured in a few cases. Consequently, the parameters of the metallic band can be established only in a few cases and these are summarized in Table 2.2.

3 The Charge Density Wave Transition and Ground State: Mean Field Theory and Some Basic Observations

Nonchalamment assis, mille couples d'amants
S'y jurent à leur aise une flamme éternelle.

A thousand couples sit in languid poses,
Exchanging pledges of undying love.

—Charles Germain De Saunt-Aubin *Sonette*

The charge density wave ground state develops in low-dimensional metals as a consequence of electron-phonon interactions. As the name suggests, the resulting ground state consists of a periodic charge density modulation accompanied by a periodic lattice distortion, both periods being determined by the Fermi wavevector k_F. Consequently both the electron and phonon spectra are strongly modified by the formation of the charge density waves. The phenomenon is usually described by discussing the behavior of a one-dimensional coupled electron-lattice system, with the electrons forming a one-dimensional electron gas, and the ions forming a linear chain. This description is adopted here.

The consequence of the electron-phonon interaction and of the divergent electronic response at $q = 2k_F$ in one dimension is a strongly renormalized phonon spectrum generally referred to as the Kohn anomaly (see Woll and Kohn, 1962 and references cited therein). By virtue of $\chi(q, T)$ this renormalization is strongly temperature-dependent, with $\omega_{\text{ren},2k_F} \to 0$ at a finite temperature (within the framework of mean field theory). This identifies a phase transition to a state where a periodic static lattice distortion and a periodically varying charge modulation with a wavelength $\lambda_0 = \pi/k_F$ develops.

For a partially filled electron band, the period λ_0 is incommensurate with the underlying lattice. The periodically varying lattice distortion leads in turn to a single particle gap at the Fermi level, turning the material into an insulator. The transition is generally referred to as the Peierls transition, since this was first suggested by Peierls (1955); the thermodynamics of the ground state have been worked out by Kuper (1955). Independently, Fröhlich (1954) suggested that the ground state can, under the influence of an applied electric field, carry an electric current. Hence the state is also often referred to as the Peierls-Fröhlich-Kuper ground state.

In this chapter we first consider the effects of electron-phonon interactions for a one-dimensional electron gas and describe the Kohn anomaly for $T > T_{CDW}^{MF}$, the mean field transition temperature. This is followed by a discussion of the state of affairs at $T = 0$, described within the framework of a weak coupling theory. As will be discussed, in the weak coupling limit, where the single particle gap is much smaller than the Fermi energy ϵ_F, the thermodynamics of the phase transition and the temperature dependence of the order parameter are the same as those of a BCS superconductor, and extensive use of the expressions worked out for the superconducting state will be made when finite temperature effects are discussed.

The most important experimental evidence for these phenomena will be discussed next, including observations of the metal-to-insulator transition which occurs at T_{CDW}, the Kohn anomaly above T_{CDW}, the single particle gap, and the periodic lattice distortion below the phase transition.

It should be stressed at this point that serious deviations from the mean field treatment are expected because of the low-dimensional character of the materials in which the formation of charge density waves have been observed and also because of the relatively short coherence lengths which result from the high transition temperatures. These effects are related to fluctuations of the order parameter and will be discussed in Chapter 5.

3.1 The Kohn Anomaly and the Peierls Transition: Mean Field Theory

In order to describe the transition to a charge density wave ground state let us consider a one-dimensional free electron gas coupled to the underlying chain of ions through electron-phonon coupling. The Hamiltonian for the electron gas is given in second quantized form as

$$\mathscr{H}_{\mathrm{el}} = \sum_k \epsilon_k a_k^\dagger a_k \tag{3.1}$$

where $a_k^\dagger$ and a_k are the creation and annihilation operators for the electron states with energy $\epsilon_k = \hbar^2 k^2/2m$. The formalism is similar if a tight binding approximation is used, however, this leads to numerical constants which are somewhat different from those which follow from the free electron approximation (Allender et al., 1974). Because spin dependent interactions are not important in the following discussion, the spin degrees of freedom are omitted and the density of states $n(\epsilon_F)$ refers to one spin direction.

The lattice vibrations are described by the Hamiltonian

$$\mathscr{H}_{\mathrm{ph}} = \sum_q \left\{ \frac{P_q P_{-q}}{2M} + \frac{M\omega_q^2}{2} Q_q Q_{-q} \right\} \tag{3.2}$$

where Q_q and P_q are the normal coordinates and conjugate state momenta of the ionic motions respectively, ω_q are the normal mode frequencies, and M is the ionic mass. With the notation

$$\begin{aligned} Q_q &= \left(\frac{\hbar}{2M\omega_q} \right)^{1/2} \left(b_q + b_{-q}^\dagger \right) \\ P_q &= \left(\frac{\hbar M \omega_q}{2} \right)^{1/2} \left(b_q^\dagger - b_{-q} \right) \end{aligned} \tag{3.3}$$

the Hamiltonian is written as

$$\mathscr{H}_{\mathrm{ph}} = \sum_q \hbar\omega_q \left(b_q^\dagger b_q + \tfrac{1}{2} \right) \tag{3.4}$$

where $b_q^\dagger$ and b_q are the creation and annihilation operators for phonons characterized by the wavevector q. In terms of these operators the lattice displacement is given by

$$u(x) = \sum_q \left(\frac{\hbar}{2NM\omega_q} \right)^{1/2} \left(b_q + b_{-q}^\dagger \right) e^{iqx} \tag{3.5}$$

where N is the number of lattice sites per unit length.

The description of the electron-phonon interaction is usually referred to as a rigid ion approximation: it is assumed that the ionic potential V at any point depends only on the distance from the center of the ion. In second quantized notation

$$\mathscr{H}_{\text{el-ph}} = \sum_{k,k',l} \langle k|V(r-l-u)|k'\rangle a_k^\dagger a_{k'} \tag{3.6}$$
$$= \sum_{k,k',l} e^{i(k'-k)(l+u)} V_{k-k'} a_k^\dagger a_{k'}$$

where $V_{k-k'}$ is the Fourier transform of the potential of a single atom $V(r)$, l gives the equilibrium lattice positions, and u is the displacement from the equilibrium position. For sufficiently small ionic displacement

$$e^{i(k'-k)u} \simeq 1 + i(k'-k)u = 1 + iN^{-1/2}(k'-k)\sum_q e^{iql} u_q. \tag{3.7}$$

The Hamiltonian, $\mathscr{H}_{\text{el-ph}}$, then has two parts. The first describes the interaction of the electrons with the underlying ions in their non-displaced positions. This term transforms the free electron spectrum into Bloch states, leading to gaps at the Brillouin zone edge. This term, will be neglected as we are interested in effects which occur at k_F, which is assumed to be away from the Brillouin zone boundaries. The second term is given by

$$\mathscr{H}_{\text{el-ph}} = iN^{1/2} \sum_{k,k',l,q} e^{i(k'-k+q)l}(k'-k)u_q V_{k-k'} a_k^\dagger a_{k'} \tag{3.8}$$
$$= iN^{1/2} \sum_{k,k'} (k'-k) u_{k-k'} V_{k-k'} a_k^\dagger a_{k'}$$

In terms of the phonon creation and annihilation operators, this interaction term becomes

$$\mathscr{H}_{\text{el-ph}} = i \sum_{k,k'} \left(\frac{\hbar}{2M\omega_{k-k'}} \right)^{1/2} \tag{3.9}$$
$$\times (k'-k) V_{k-k'} \left(b_{k'-k}^\dagger + b_{k-k'}\right) a_k^\dagger a_{k'}$$
$$= \sum_{k,q} g_q \left(b_{-q}^\dagger + b_q\right) a_{k+q}^\dagger a_k$$

where the electron-phonon coupling constant is

$$g_q = i\left(\frac{\hbar}{2M\omega_q}\right)^{1/2} |q| V_q . \tag{3.10}$$

With all the terms collected, the Hamiltonian, often called the Fröhlich Hamiltonian (Fröhlich, 1954), is written as

$$\mathscr{H} = \sum_k \epsilon_k a_k^\dagger a_k + \sum_q \hbar\omega_q b_q^\dagger b_q + \sum_{k,q} g_q a_{k+q}^\dagger a_k \left(b_{-q}^\dagger + b_q\right). \tag{3.11}$$

In the case of a 1D electron gas, for which the dispersion relation is described by $\epsilon_k = \hbar v_F(k - k_F) = \pm v_F \delta k$ near $\pm k_F$ (see Fig. 2.1), (with the energy measured from the Fermi energy ϵ_F) the coupled electron-phonon system is unstable, and this instability has fundamental consequences for both the lattice and the electron gas.

The effect of the electron-phonon interaction on the lattice vibrations can be described by establishing the equation of motion of the normal coordinates. For small amplitude displacements

$$\hbar^2 \ddot{Q}_q = -\left[\left[Q_q, \mathscr{H}\right], \mathscr{H}\right] \tag{3.12}$$

where $\ddot{Q}_q$ refers to the second time derivative of the coordinate. The above equation, utilizing the commutation relations

$$\left[Q_q, P_{q'}\right] = i\hbar\delta_{q,q'} \tag{3.13}$$

becomes

$$\ddot{Q}_q = -\omega_q^2 Q_q - g\left(\frac{2\omega_q}{M\hbar}\right)^{1/2} \rho_q \tag{3.14}$$

where we have assumed that g is independent of q, and $\rho_q = \Sigma_k a_{k+q}^\dagger a_k$ is the qth component of the electronic density. The second term on the right hand side is an effective force associated with the lattice dynamics and arises as a consequence of electron-phonon interactions. The ionic potential $g(2M\omega_q/\hbar)^{1/2} Q_q$ gives, through Eq. (1.6), a density fluctuation

$$\rho_q = \chi(q,T)\, g\left(\frac{2M\omega_q}{\hbar}\right)^{1/2} Q_q \tag{3.15}$$

and with this mean field approximation Eq. (3.14) reads

$$\ddot{Q}_q = -\left[\omega_q^2 + \frac{2g^2\omega_q}{M\hbar}\chi(q,T)\right]Q_q. \tag{3.16}$$

The above equation of motion gives a renormalized phonon frequency

$$\omega_{\mathrm{ren},q}^2 = \omega_q^2 + \frac{2g^2\omega_q}{\hbar}\chi(q,T). \tag{3.17}$$

As discussed in Chapter 1, for a one-dimensional electron gas $\chi(q,T)$ has its maximum value at $q = 2k_F$. Consequently, the reduction, or softening, of the phonon frequencies will be most significant at these wavevectors. The phonon frequency for $q = 2k_F$ becomes

$$\omega_{\mathrm{ren},2k_F}^2 = \omega_{2k_F}^2 - \frac{2g^2 n(\epsilon_F)\omega_{2k_F}}{\hbar} ln\left(\frac{1.14\epsilon_0}{k_B T}\right). \tag{3.18}$$

With decreasing temperature, the renormalized phonon frequency goes to zero and this defines a transition temperature, when a frozen-in distortion occurs. From Eq. (3.18)

$$k_B T_{\mathrm{CDW}}^{\mathrm{MF}} = 1.14\epsilon_0 e^{-1/\lambda} \tag{3.19}$$

where λ is the dimensionless electron-phonon coupling constant

$$\lambda = \frac{g^2 n(\epsilon_F)}{\hbar\omega_{2k_F}} = g' n(\epsilon_F). \tag{3.20}$$

Close to the transition temperature a straightforward expansion gives the following temperature dependence

$$\omega_{\mathrm{ren},2k_F} = \omega_{2k_F}\left(\frac{T - T_{\mathrm{CDW}}^{\mathrm{MF}}}{T_{\mathrm{CDW}}^{\mathrm{MF}}}\right)^{1/2} \tag{3.21}$$

The phonon dispersion relation $\omega_{\mathrm{ren},q}$, as determined by Eq. (3.17), is shown in Fig. 3.1 at various temperatures above the mean field transition temperature. This dispersion relation can be calculated by using the expression for $\chi(q = 2k_F, T)$ as given by Eq. (1.10). The detailed form, as expected, also depends on the phonon dispersion relation, and in Fig. 3.1 the situation appropri-

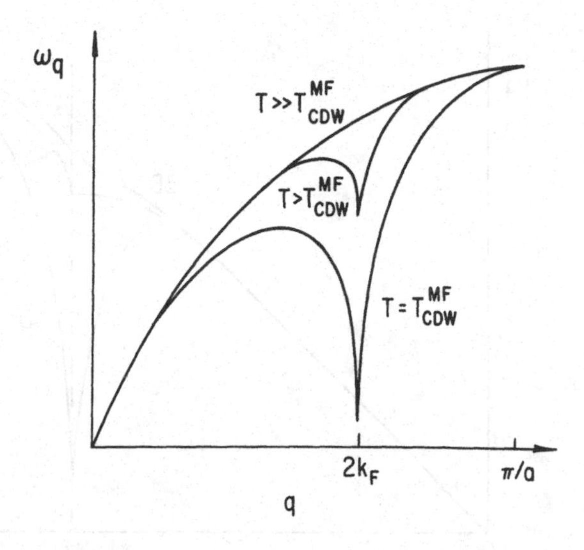

Figure 3.1. Acoustic phonon dispersion relation of a one-dimensional metal at various temperatures above the mean field transition temperature.

ate for the renormalization of an acoustic phonon branch at various temperatures is displayed.

As discussed earlier, the phase transition is defined by the temperature where $\omega_{\mathrm{ren},2k_F} \to 0$ and is due to the strongly divergent response function of the 1D electron gas. The renormalization of the phonon frequencies also occurs in higher dimensions, however the reduction of the phonon frequencies is less significant. Expressions for $\chi(q,T)$ appropriate for a 2D and 3D electron gas can also be used to evaluate the renormalized phonon frequencies and these are shown, together with the 1D case, in Fig. 3.2. For the higher dimensions, the temperature dependence of $\chi(q,T)$ is weak, and for small electron-phonon coupling constants $\omega_{\mathrm{ren},q}$ remains finite at $T=0$ and there is no phase transition.

Below the phase transition, the renormalized phonon frequency is zero, indicating a "frozen-in" lattice distortion, that is, a macroscopically occupied phonon mode with nonvanishing expectation values $\langle b_{2k_F}\rangle = \langle b^{\dagger}_{-2k_F}\rangle$. The order parameter is defined as

$$|\Delta| e^{i\phi} = g\left(\langle b_{2k_F}\rangle + \langle b^{\dagger}_{-2k_F}\rangle\right) \tag{3.22}$$

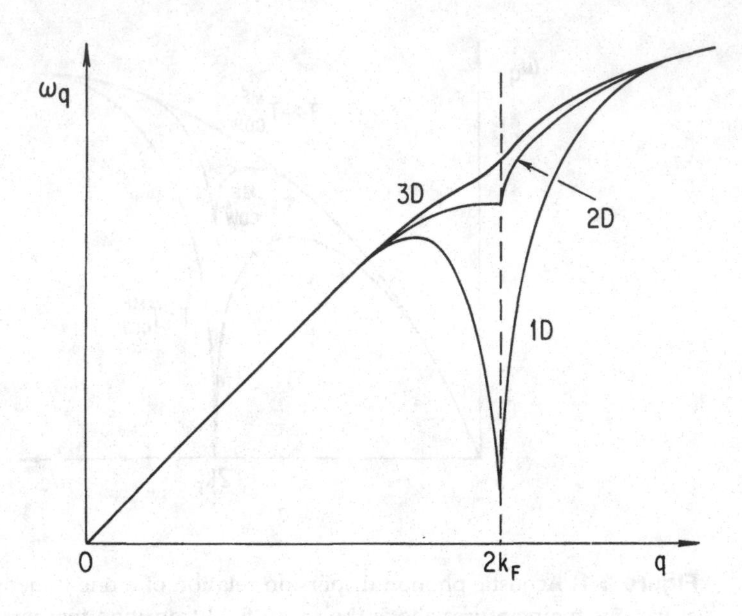

Figure 3.2. Acoustic phonon dispersion relations of one-, two-, and three-dimensional metals.

and from Eq. (3.5) the lattice displacement is given by

$$\begin{aligned}\langle u(x)\rangle &= \left(\frac{\hbar}{2NM\omega_{2k_F}}\right)^{1/2}\left\{i(\langle b_{2k_F}\rangle + \langle b^{+}_{-2k_F}\rangle)e^{i2k_Fx} + c.c.\right\}\\ &= \left(\frac{\hbar}{2NM\omega_{2k_F}}\right)^{1/2}\frac{2|\Delta|}{g}\cos(2k_Fx+\phi) \qquad (3.23)\\ &= \Delta u\cos(2k_Fx+\phi)\end{aligned}$$

with

$$\Delta u = \left(\frac{2\hbar}{NM\omega_{2k_F}}\right)^{1/2}\frac{|\Delta|}{g}.$$

With this mean field approximation Eq. (3.11) becomes

$$\mathscr{H} = \sum_k \epsilon_k a^{\dagger}_k a_k + \sum_q \hbar\omega_q\langle b^{\dagger}_q b_q\rangle + \sum_{k,q} g a^{\dagger}_{k+q} a_k\langle b^{\dagger}_{-q} + b_q\rangle \qquad (3.24)$$

where $q = \pm 2k_F$. As $\langle b_{2k_F} \rangle = \langle b^{\dagger}_{-2k_F} \rangle$, the Hamiltonian becomes

$$\mathscr{H} = \sum_k \epsilon_k a_k^{\dagger} a_k + 2g \sum_k \left[a^{\dagger}_{k+2k_F} a_k \langle b^{\dagger}_{-2k_F} \rangle + a^{\dagger}_{k-2k_F} a_k \langle b_{-2k_F} \rangle \right] + 2\hbar\omega_{2k_F} \langle b_{2k_F} \rangle^2. \tag{3.25}$$

With the order parameter defined in (3.22) the electronic part becomes

$$\mathscr{H}_{\text{el}} = \sum_k \left[\epsilon_k a_k^{\dagger} a_k + |\Delta| e^{i\phi} a^{\dagger}_{k+2k_F} a_k + |\Delta| e^{-i\phi} a^{\dagger}_{k-2k_F} a_k \right]. \tag{3.26}$$

In the spirit of the nearly free electron approximation we consider only states near the Fermi level, and states near $+k_F$ and near $-k_F$ are labelled by the subscripts 1 and 2 respectively. With this notation and using the dispersion relation $\varepsilon_k = \hbar v_F (k - k_F)$, Eq. (3.26) becomes

$$\mathscr{H} = \sum_k \left[\epsilon_k \left(a^{\dagger}_{1,k} a_{1,k} - a^{\dagger}_{2,k} a_{2,k} \right) + |\Delta| e^{i\phi} a^{\dagger}_{1,k} a_{2,k} + |\Delta| e^{-i\phi} a^{\dagger}_{2,k} a_{1,k} \right]. \tag{3.27}$$

Here and in the following, k is measured relative to the Fermi wavevector k_F. The Hamiltonian can be diagonalized by a canonical transformation. We define a new set of operators by

$$\gamma_{1,k} = U_k a_{1,k} - V_k^* a_{2,k} = U_k e^{-i\phi/2} a_{1,k} - V_k e^{i\phi/2} a_{2,k} \tag{3.28a}$$

$$\gamma_{2,k} = V_k a_{1,k} + U_k^* a_{2,k} = V_k e^{-i\phi/2} a_{1,k} + U_k e^{i\phi/2} a_{2,k} \tag{3.28b}$$

with the constraint that $U_k^2 + V_k^2 = 1$. After some algebra, Eq. (3.27) in terms of $\gamma_{1,k}$ and $\gamma_{2,k}$, becomes

$$\mathscr{H} = \sum_k \left[\epsilon_k (U_k^2 - V_k^2) - 2|\Delta| U_k V_k \right] \left(\gamma^{\dagger}_{1,k} \gamma_{1,k} - \gamma^{\dagger}_{2,k} \gamma_{2,k} \right) + \left[2\epsilon_k U_k V_k + |\Delta| (U_k^2 - V_k^2) \right] \left(\gamma^{\dagger}_{1,k} \gamma_{2,k} + \gamma^{\dagger}_{2,k} \gamma_{1,k} \right). \tag{3.29}$$

The Hamiltonian is diagonalized if the coefficients in front of the off-diagonal terms are zero; for example,

$$2\epsilon_k U_k V_k + |\Delta|(U_k^2 - V_k^2) = 0. \tag{3.30}$$

The condition $U_k^2 + V_k^2 = 1$ is satisfied if we choose

$$V_k = \cos\left(\frac{\theta_k}{2}\right) \tag{3.31a}$$

$$U_k = \sin\left(\frac{\theta_k}{2}\right) \tag{3.31b}$$

and then (3.30) becomes

$$\tan\theta_k = -\frac{|\Delta|}{\epsilon_k}. \tag{3.32}$$

From (3.32) and (3.31) we obtain

$$V_k^2 = \frac{1}{2}\left(1 - \frac{\epsilon_k}{\left(\epsilon_k^2 + \Delta^2\right)^{1/2}}\right) = \frac{1}{2}\left(1 + \frac{\epsilon_k}{E_k}\right) \tag{3.33a}$$

$$V_k^2 = \frac{1}{2}\left(1 - \frac{\epsilon_k}{E_k}\right) \tag{3.33b}$$

with

$$E_k = \epsilon_F + \operatorname{sign}(k - k_F)\left[\hbar^2 v_F^2 (k - k_F)^2 + \Delta^2\right]^{1/2}. \tag{3.34}$$

Upon substitution of the expressions for U_k and V_k, Eq. (3.29) becomes

$$\mathscr{H} = \sum_k E_k\left(\gamma_{1,k}^\dagger \gamma_{1,k} + \gamma_{2,k}^\dagger \gamma_{2k}\right) + \frac{\hbar\omega_{2k_F}\Delta^2}{2g^2}. \tag{3.35}$$

Instead of the linear dispersion relation $\epsilon_k - \epsilon_F = \hbar v_F(k - k_F)$ appropriate for the metallic state the spectrum of excitations develops a gap in the charge density wave state, and the dispersion relation is shown in Fig. 3.3. The density of states is obtained from the condition

$$N_{\mathrm{CDW}}(E)\, dE = N_e(\epsilon)\, d\epsilon = N_e\, d\epsilon \tag{3.36}$$

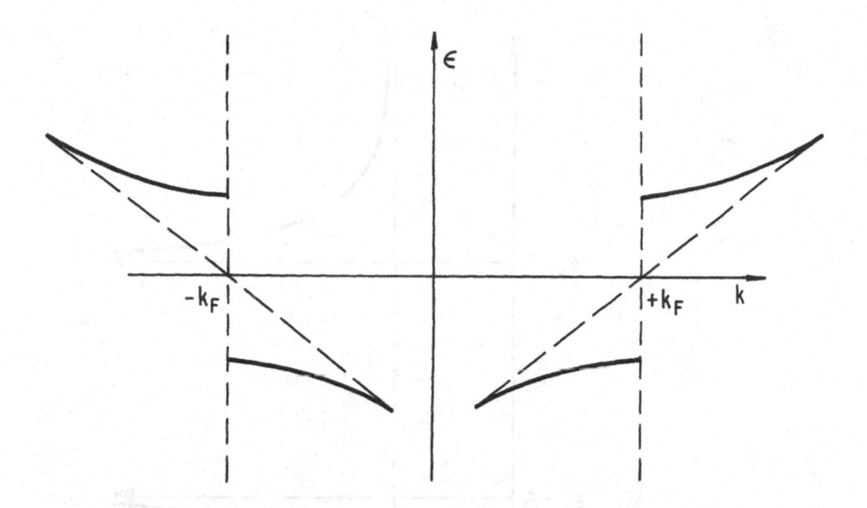

Figure 3.3. The dispersion relation for ϵ_k (dashed line) and E_k (full line) near the Fermi wavevectors $\pm k_F$.

assuming that the density of states in the metallic state, $N_e(\epsilon) = N_e$ is independent of the energy. With this assumption

$$\frac{N_{\mathrm{CDW}}(E)}{N_e} = \frac{d\epsilon}{dE} = \begin{cases} 0 & |E| < \Delta \\ \dfrac{E}{(E^2 - \Delta^2)^{1/2}} & |E| > \Delta \end{cases} \tag{3.37}$$

The density of states diverges as the gap is approached from below ($E < \epsilon_F$) or from above ($E > \epsilon_F$). The total number of states is the same in the normal and charge density wave states, and the increase of $N_{\mathrm{CDW}}(E)$ near Δ is due to the removal of the density of states within the region of the single particle gap. Both N_{CDW} and $N_e(\epsilon)$, which is assumed to be constant, are shown in Fig. 3.4.

The magnitude of the gap is obtained by minimizing the energy of Eq. (3.35). The opening of the gap leads to the lowering of the electronic energy, this is given by

$$E_{\mathrm{el}} = \sum_k (-E_k + v_F k) = n(\epsilon_F) \int_0^{\epsilon_F} \left(\epsilon - (\epsilon^2 + \Delta^2)^{1/2}\right) d\epsilon \tag{3.38}$$

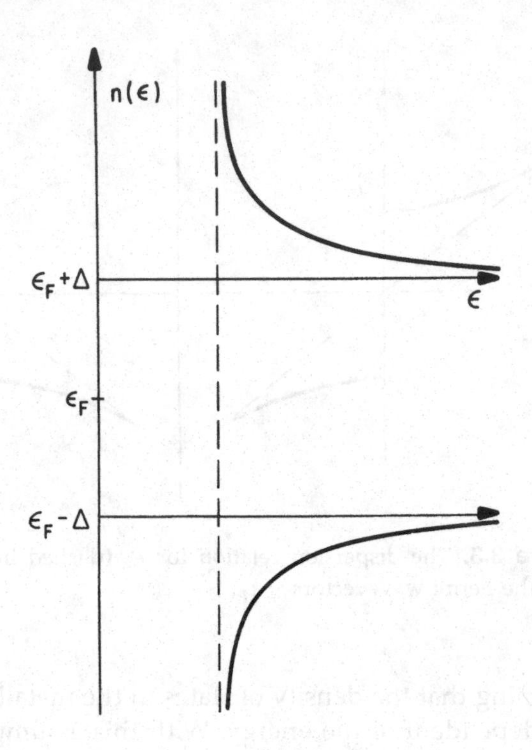

Figure 3.4. The energy dependence of the density of states near $\pm k_F$ in the metallic (dashed line) and in the charge density wave (full line) states.

which after some algebra becomes

$$E_{\text{el}} = \frac{n(\epsilon_F)}{2}\left\{\epsilon_F^2 - \left[\epsilon_F\left(\epsilon_F^2+\Delta^2\right)^{1/2} + \Delta^2 \log\frac{\epsilon_F + \left(\epsilon_F^2-\Delta^2\right)^{1/2}}{\Delta}\right]\right\}. \tag{3.39}$$

In the weak coupling limit ($\epsilon_F \gg \Delta$) the expansion of the log term gives

$$E_{\text{el}} = n(\epsilon_F)\left[-\frac{\Delta^2}{2} - \Delta^2 \log\left(\frac{2\epsilon_F}{\Delta}\right)\right] + O\left(\frac{\Delta}{\epsilon_F}\right)\Bigg]. \tag{3.40}$$

The lattice distortion leads to an increase in the elastic energy, given by

$$E_{\text{latt}} = \frac{N}{2} M \omega_{2k_F}^2 \langle u(x) \rangle^2 \tag{3.41}$$

which, with Eq. (3.23), becomes

$$E_{\text{latt}} = \frac{\hbar \omega_{2k_F} \Delta^2}{2g^2} = \frac{\Delta^2 n(\epsilon_F)}{\lambda}, \tag{3.42}$$

where λ is the dimensionless coupling constant defined in Eq. (3.20). The total energy change is given by the two terms

$$E_{\text{tot}} = E_{\text{el}} + E_{\text{latt}} = n(\epsilon_F)\left[-\frac{\Delta^2}{2} - \Delta^2 \log\left(\frac{2\epsilon_F}{\Delta}\right) + \frac{\Delta^2}{2\lambda} \right]. \tag{3.43}$$

Straightforward minimization of the total energy leads, for $\lambda << 1$, to

$$\Delta = 2\epsilon_F e^{-1/\lambda} \tag{3.44}$$

and to the condensation energy

$$E_{\text{con}} = E_{\text{normal}} - E_{\text{CDW}} = \frac{n(\epsilon_F)}{2} \Delta^2. \tag{3.45}$$

The positive sign indicates that the CDW ground state has, in 1D, lower energy than the normal state. A comparison with Eq. (3.19) leads to

$$2\Delta = 3.52 k_B T_{\text{CDW}}^{\text{MF}}, \tag{3.46}$$

the well known BCS relation between the zero temperature gap and the transition temperature.

The ground state wave function

$$\begin{aligned} |\phi_0\rangle &= \left(\prod_{|k|<k_F} \gamma_{1,k}^{\dagger} \gamma_{2,k}^{\dagger} \right) |0\rangle \\ &= \prod_{|k|<k_F} \left(U_k e^{i\phi/2} a_{1,k}^{\dagger} - V_k e^{-i\phi/2} a_{2,k}^{\dagger} \right) \\ &\quad \times \left(V_k e^{i\phi/2} a_{1,k}^{\dagger} + U_k e^{-i\phi/2} a_{2,k}^{\dagger} \right) |0\rangle \end{aligned} \tag{3.47}$$

where $|0\rangle$ represents the vacuum. The electronic density is

$$\rho(x) = \langle \phi_0 | \Psi^*(x) \Psi(x) | \phi_0 \rangle \tag{3.48}$$

where

$$\Psi(x) = \sum_k \left(a_{1,k} e^{ik_l x} + a_{2,k} e^{-ik_l x}\right). \tag{3.49}$$

Substituting the operators $\gamma_{1,k}$ and $\gamma_{2,k}$ from Eq. (3.28) and using the relations

$$\langle \phi_0 | \gamma^{\dagger}_{2,k} \gamma_{2,k} | \phi_0 \rangle = \langle \phi_0 | \gamma^{\dagger}_{1,k} \gamma_{1,k} | \phi_0 \rangle = 1 \tag{3.50a}$$

$$\langle \phi_0 | \gamma^{\dagger}_{1,k} \gamma_{2,k} | \phi_0 \rangle = \langle \phi_0 | \gamma^{\dagger}_{2,k} \gamma_{1,k} | \phi_0 \rangle = 0 \tag{3.50b}$$

straightforward algebra gives

$$\rho(x) = \sum_k \left[U_k^2 + V_k^2 - U_k V_k e^{i(2k_l x + \phi)} - V_k U_k e^{-i(2k_F x + \phi)} \right] \tag{3.51}$$

where U_k and V_k are given in Eq. (3.33). Integration over the occupied states leads, in the weak coupling limit, to a periodic charge density variation:

$$\rho(x) = \rho_0 \left[1 + \frac{\Delta}{\hbar v_F k_F \lambda} \cos(2k_F x + \phi) \right] \tag{3.52}$$

where ρ_0 is the (constant) electronic density in the metallic state. The dispersion relation, the electronic density, and the equilibrium lattice positions are shown, both above T^{MF}_{CDW} and at $T = 0$ in Fig. 3.5. (For the purpose of demonstration of what happens, a half-filled band is shown on the figure. For this special case, $\lambda_0 = 2a$ and the density wave is not incommensurate with the underlying lattice. These so-called commensurability effects will be discussed in Chapter 7.) The ground state exhibits a periodic modulation of the charge density and a lattice distortion (hence the name charge density wave state) and the single particle excitations have a gap Δ at the Fermi level. This turns the material into an insulator (if the condensate does not contribute to the electrical conduction) upon the formation of the charge density wave ground state. The period $\lambda_0 = \pi / k_F$ is incommensurate with the underlying lattice for an arbitrary band filling. In this case, the condensation energy, Eq. (3.45), is independent of the phase variable ϕ and this has important consequences both in terms of the collective excitations and of the response of the charge density wave to external *dc* and *ac* electromagnetic fields. Commensurability effects which lead to a position-dependent

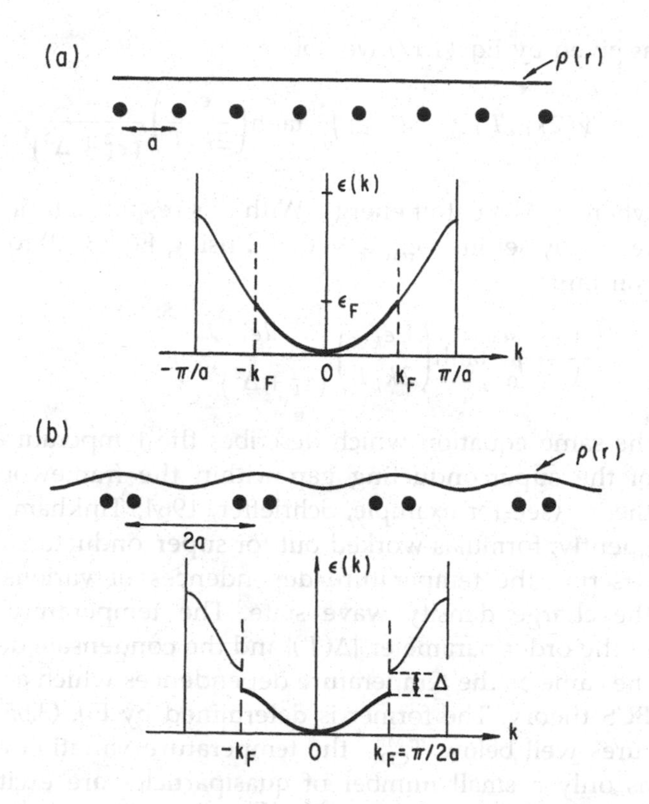

Figure 3.5. The single particle band, electron density, and lattice distortion in the metallic state above T_{CDW} and in the charge density wave state at $T = 0$. The figure is appropriate for a half-filled band.

condensation energy are neglected here; they will be discussed in Chapter 7. The above discussion of the charge density wave state has been at temperature $T = 0$. At finite temperature, thermally induced transitions across the gap lead to the screening of the electron-phonon interaction, to the reduction of the energy gain by the gap, and eventually to a phase transition. This can be discussed considering Eq. (3.17) which reads

$$\omega_{2k_F} = -\frac{2g^2}{\hbar}\chi(2k_F, T) \tag{3.53}$$

for $\omega_{\text{ren},2k_F} = 0$. Because of the single particle gap, the response function $\chi(q, T)$ is different from that of a metallic one-dimensional band. Using arguments similar to those which lead to $\chi(q)$

as given by Eq. (1.17) we obtain

$$\chi(2k_F, T) = -n(\epsilon_F)\int_0^{\epsilon_0} \tanh\left(\frac{\epsilon_k}{2k_BT}\right)\frac{d\epsilon_k}{(\epsilon_k^2+\Delta^2)^{1/2}} \quad (3.54)$$

where ϵ_0 is a cutoff energy. With this response function, Eq. (3.18) reads (by setting $\omega_{\text{ren},2k_F} = 0$ and using Eq. (3.20) for the coupling constant)

$$\frac{1}{\lambda} = \int_0^{\epsilon_0} \tanh\left(\frac{\epsilon_k}{2k_BT}\right)\frac{d\epsilon_k}{(\epsilon_k^2+\Delta^2)^{1/2}}, \quad (3.55)$$

the same equation which describes the temperature dependence of the superconducting gap within the framework of the BCS theory (see for example, Schrieffer, 1964; Tinkham, 1975). Consequently, formulas worked out for superconductors can be used to describe the temperature dependences of various quantities in the charge density wave state. The temperature dependences of the order parameter $|\Delta(T)|$ and the condensate density $f(T)$ are the same as the temperature dependences which are given by the BCS theory. The former is determined by Eq. (3.55). At temperatures well below $T_{\text{CDW}}^{\text{MF}}$ the temperature variation of $|\Delta|$ is weak, as only a small number of quasiparticles are excited across the single particle gap. Close to $T_{\text{CDW}}^{\text{MF}}$, where the gap approaches zero as $T \to T_{\text{CDW}}^{\text{MF}}$ from below, Eq. (3.19) reduces to

$$\frac{|\Delta(T)|}{|\Delta(0)|} = 1.74\left(1 - \frac{T}{T_{\text{CDW}}^{\text{MF}}}\right)^{1/2}. \quad (3.56)$$

The condensate density depends on the frequency and wavevector (Rice et al., 1979; Maki and Virosztek, 1990). In the static ($\omega \ll v_Fq$) limit it is given by the relation

$$f_s(T) = |\Delta|^2 \pi T \sum_{\omega_n} (\omega_n^2 + \Delta^2)^{-3/2} \quad (3.57a)$$

and in the opposite, so-called dynamic limit ($\omega \gg v_Fq$)

$$f_d(T) = |\Delta|^2 \int_\Delta^\infty dE \frac{\tanh(E/2k_BT)}{(E^2-\Delta^2)^{1/2}E^2}. \quad (3.57b)$$

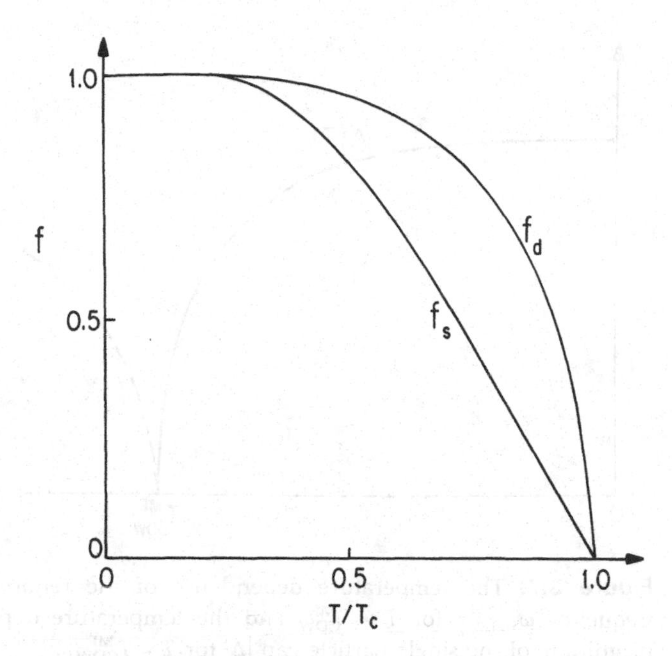

Figure 3.6. Temperature dependence of the static and dynamic condensate density as given by Eqs. (3.57a) and (3.57b).

As $T \to 0$, both f_s and $f_d \to 1$, and close to the transition temperature

$$f_s(T) \simeq 2\left(1 - \frac{T}{T_{\text{CDW}}^{\text{MF}}}\right)$$
$$f_d(T) \simeq \frac{\pi}{4}\frac{|\Delta(T)|}{T_{\text{CDW}}^{\text{MF}}}. \tag{3.58}$$

The temperature dependences of the two parameters are together with $\Delta(T)$ and $\omega_{\text{ren}}(T)$ displayed in Fig. 3.6 and Fig. 3.7.

The behaviors of the various thermodynamic quantities are the same as those of a superconductor. The specific heat is exponentially small at low temperatures, while at T_{CDW} there is a finite discontinuity, given by

$$\Delta C = n(\epsilon_F)\left(-\frac{d\Delta^2}{dT}\right) \tag{3.59}$$

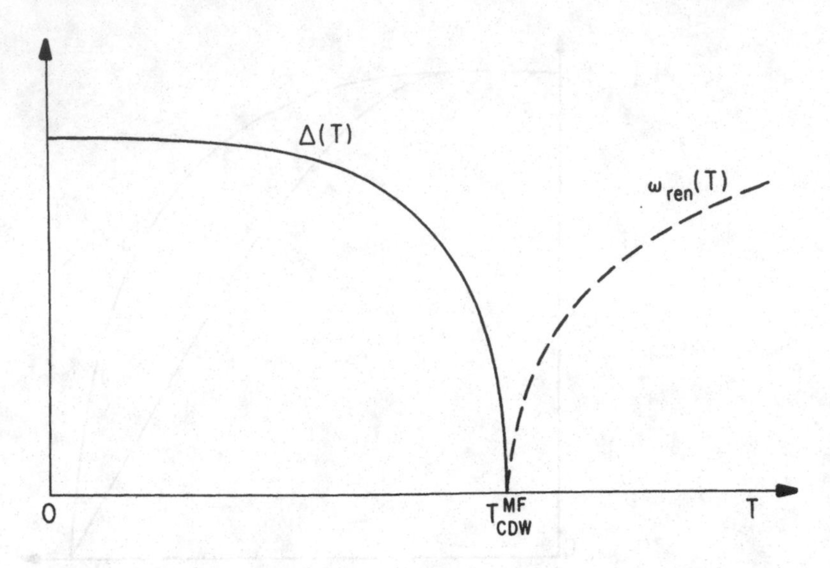

Figure 3.7. The temperature dependence of the renormalized phonon frequency $\omega_{\text{ren},2k_F}$ for $T > T_{\text{CDW}}^{\text{MF}}$ and the temperature dependence of the magnitude of the single particle gap $|\Delta|$ for $T < T_{\text{CDW}}^{\text{MF}}$.

leading to

$$\Delta C = \frac{\pi^2 T_{\text{CDW}}^{\text{MF}} n(\epsilon_F) k_B^2}{7\zeta(3)} = 1.438\gamma_e T_{\text{CDW}}^{\text{MF}} k_B^2 n(\epsilon_F) \tag{3.60}$$

in the weak coupling limit. Here $\gamma_e = 2\pi^2/3n(\epsilon_F)k_B^2$ is the linear specific heat coefficient of the electron gas and $\zeta(3) = 1.202$ is the Riemann zeta function.

As usual, the behavior near the transition temperature can be discussed using the Landau theory of phase transitions. The functional form of the free energy is chosen to be

$$F = F(0) + n(\epsilon_F) \int dx \{a|\Delta|^2 + b|\Delta|^4 + \cdots\} \tag{3.61}$$

With $a = a_1(T - T_{\text{CDW}})$ and $b = \text{const}$, the temperature dependent order parameter is obtained from the equilibrium condition

$$\frac{\partial F}{\partial \Delta} = 0.$$

This, together with Eq. (3.61), leads to $|\Delta| = 0$ above the transition temperature, while for $T < T_{\text{CDW}}^{\text{MF}}$

$$|\Delta(T)| = \left(\frac{a_1}{2b}\right)^{1/2} (T_{\text{CDW}}^{\text{MF}} - T)^{1/2} \tag{3.62a}$$

which is identical to Eq. (3.56) if

$$\left(\frac{a_1}{8b}\right)^{1/2} = 1.74|\Delta(T=0)|. \tag{3.62b}$$

The specific heat jump at the transition is

$$\Delta C = \frac{a_1^2 n(\epsilon_F)}{2bT_{\text{CDW}}^{\text{MF}}} \tag{3.63a}$$

which, by comparison to Eq. (3.60), leads to the following relation

$$\frac{a_1^2}{2bT_{\text{CDW}}^{\text{MF}}} = \frac{8\pi^2 T_{\text{CDW}}^{\text{MF}}}{7\zeta(3)}. \tag{3.63b}$$

From Eqs. (3.62a) and (3.63a) the coefficients of the Ginzburg-Landau theory are given in terms of the microscopic parameters as

$$a = \frac{T}{T_{\text{CDW}}^{\text{MF}}}; \qquad a_1 = \frac{1}{T_{\text{CDW}}^{\text{MF}}} \tag{3.64}$$

and

$$b = \frac{7\zeta(3)}{16\pi^2 (T_{\text{CDW}}^{\text{MF}})^2} \tag{3.65}$$

near the mean field transition temperature.

Above $T_{\text{CDW}}^{\text{MF}}$, the minimum of the free energy is at $|\Delta| = 0$ and the slope, with respect to the fluctuating order parameter, is given by

$$\left.\frac{\partial^2 F}{\partial \Delta^2}\right|_{|\Delta|=0} = a + 3b|\Delta|^2 \simeq a_1(T - T_{\text{CDW}}^{\text{MF}}). \tag{3.66}$$

This slope corresponds to a restoring force for the oscillation of the order parameter, and the oscillation frequency is given, near T_{CDW} by

$$\omega_0 = \left(\frac{k}{M}\right)^{1/2} = \left(\frac{a_1}{M}\right)^{1/2} (T - T_{\text{CDW}})^{1/2} \tag{3.67}$$

in agreement with the temperature dependence of the frequency of the soft phonon mode at $q=2k_F$, which was evaluated above in Eq. (3.21).

Some comments on the relevant cutoff frequency and the effects of higher dimensionality are in order here. For charge density waves, the energy gaps develop in response to a static distortion of the underlying lattice and relaxation effects do not play a role. Consequently, the entire electron spectrum is influenced by the electron-phonon interaction and the relevant cutoff energy is the Fermi energy ϵ_F. This is in contrast to the superconducting ground state, where retardation of the electron-phonon interaction is essential and leads in turn to a cutoff frequency on the order of ω_D, the Debye frequency. The difference between the two energy scales, with ϵ_F usually significantly larger than $\hbar\omega_D$, leads in general to values of T_{CDW}^{MF} which exceed the typical superconducting transition temperatures.

The singularity of the density of states at the single particle gap, as given by Eq. (3.37), is due to the 1D character of the electronic spectrum. Higher dimensional effects, which can be described by a quasi-2D or quasi-3D band structure, lead to (along with a modification of the response function $\chi(\vec{q})$) the removal of these singularities at $q=2k_F$, and thus to changes in the electronic energies upon the formation of the single particle gap (Huang and Maki, 1989). Consequently, the condition given by Eq. (3.46) is also affected, leading to a single particle gap and a transition temperature which is different from that which is appropriate for the strictly 1D case.

3.2 Single Particle Transitions: Tunneling and Coherence Factors

The single particle gap and density of states which develop below T_{CDW}^{MF} as a consequence of charge density wave formation are, for the strictly one-dimensional case, identical to the spectrum of single particle excitations for the singlet superconductor. Although the ground states are different; because of this similarity, the formalism which has been developed for single particle tunneling in superconductors can be adopted to describe the occurrence of tunneling in the charge density wave ground state. To date, tunneling has only been observed between a normal metal and materials with a charge density wave state, therefore expressions appropriate for this case will be reproduced here. In terms of the standard semiconductor model (also applied for superconductors; see for example Tinkham, 1975), the CDW ground state is viewed as a one-dimensional semiconductor and the density of states is given by Eq. (3.37). With the normal metal represented by an energy independent density of states N_e, the tunneling current is given by

$$I_{n,\mathrm{CDW}} = A\frac{G_{nn}}{e}\int_{-\infty}^{\infty}\frac{N_{\mathrm{CDW}}(E)}{N_e}[f(E)-f(E+eV)]\,dE \tag{3.68}$$

where G_{nn} is the tunneling conductance for normal metal to normal metal tunneling and $N_{\mathrm{CDW}}(E)$ is the density of states in the charge density wave state. As for a superconductor, at $T=0$ there is no tunneling current as long as $e|V|<|\Delta|$; for nonzero temperatures an (exponentially) small tunneling current is expected for voltages less than the single particle gap. The differential conductance dI/dV is given, in analogy to a superconductor by

$$G_{n,\mathrm{CDW}}\bigg|_{T=0} = \frac{dI_{n,\mathrm{CDW}}}{dV}\bigg|_{T=0} = G_{nn}\frac{N_{\mathrm{CDW}}e|V|}{N_e} \tag{3.69}$$

and is, at temperatures $k_BT<|\Delta|$, directly proportional to the density of states.

In addition to normal metal-charge density wave tunneling, it is expected that other tunneling phenomena, such as charge density wave-charge density wave tunneling, and phonon structures may also be observed; and the formalism applied for superconductors may be adopted. This of course does not hold for tunneling involving coherent pairs, but it has also been suggested that effects similar to Josephson tunneling may occur as well (Artemenko and Volkov, 1983).

Single electron transitions, induced by external perturbations such as electromagnetic radiation, ultrasound, or nuclear relaxation are determined by coherence factors, in addition to density of state effects. Such coherence factors arise when the transition probabilities are calculated. Because the superconducting and charge density wave ground states are invariant under different symmetry operations, the coherence factors will also be different in the two cases.

We write the external perturbation which acts on the single particle states as

$$\mathscr{H}_1 = \sum_{k,k'} B_{k',k} a^{\dagger}_{k'} a_k = \sum_{k,k'} B_{k',k}\left(a^{\dagger}_{1,k'} a_{1,k} + a^{\dagger}_{2,k'} a_{2,k}\right) \quad (3.70)$$

where the electron spin has been neglected. Here $B_{k',k}$ are the matrix elements which connect the usual one-electron states in normal metals. Each term is independent, and $|B_{k',k}|^2$ is proportional to the transition probability.

In the charge density wave state, the transition is written in terms of the quasiparticle states γ as

$$\begin{aligned} a^{\dagger}_{1,k'} a_{1,k} &= \left(U^*_{k'} \gamma^{\dagger}_{1,k'} - V^*_{k'} \gamma_{2,k'}\right)\left(U_k \gamma_{1,k} - V_k \gamma^{\dagger}_{2,k}\right) \quad (3.71a)\\ &= U^*_{k'} U_k \gamma^{\dagger}_{1,k'} \gamma_{1,k} - V^*_{k'} V_k \gamma^{\dagger}_{2,k} \gamma_{2,k'}\\ &\quad - U^*_{k'} V_k \gamma^{\dagger}_{1,k'} \gamma^{\dagger}_{2,k} - V^*_{k'} U_k \gamma_{2,k'} \gamma_{1,k}\\ a^{\dagger}_{2,k'} a_{2,k} &= \left(V_{k'} \gamma^{\dagger}_{1,k'} + U_{k'} \gamma_{2,k'}\right)\left(V^*_k \gamma_{1,k} + U^*_k \gamma^{\dagger}_{2,k}\right) \quad (3.71b)\\ &= V_{k'} V^*_k \gamma^{\dagger}_{1,k'} \gamma_{1,k} - U_{k'} U^*_k \gamma^{\dagger}_{2,k} \gamma_{2,k}\\ &\quad + V_{k'} U^*_k \gamma^{\dagger}_{1,k'} \gamma^{\dagger}_{2,k} + U_{k'} V^*_k \gamma_{2,k'} \gamma_{1,k}. \end{aligned}$$

Thus the transitions $a^{\dagger}_{1,k'} a_{1,k}$ and $a^{\dagger}_{2,k'} a_{2,k}$ connect the same quasiparticle states. Consequently, the transition probabilities of these transitions are not independent and the matrix elements must either be the same, or differ only in sign.

The first two terms on the right hand sides correspond to the scattering of quasiparticles by the external perturbation, these processes are important for energies significantly smaller than the single particle gap. The last two terms correspond to the creation or annihilation of two quasiparticles and such processes occur if the frequency of the external perturbation exceeds the gap frequency $\omega_g = 2\Delta/\hbar$. These processes are shown in Fig. 3.8.

The subscripts 1 and 2 refer to quasiparticles with $k > 0$ and $k < 0$ respectively, and the transition matrix element depends on whether the external perturbation is symmetric or antisymmetric when $k \to k - 2k_F$. The former is the case for ultrasonic attenuation (case I) as the interaction is between the one electron states and the scalar deformation potential of the ions. Then the transi-

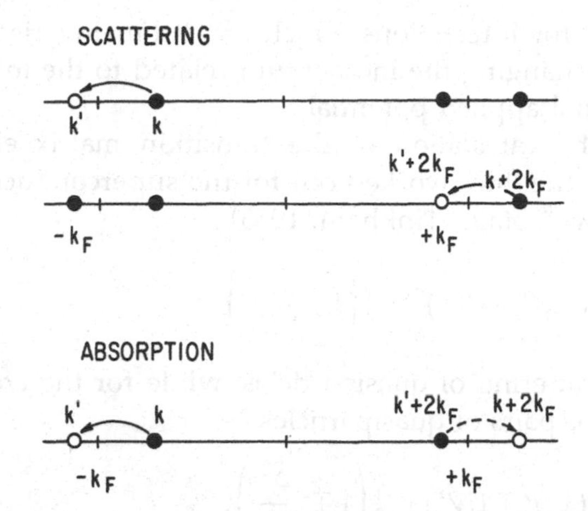

Figure 3.8. Scattering of single particles and creation of pairs of particles in the CDW state. In the upper part of the figure two scattering processes, both involving the same change of momentum $\Delta k = k' - k$ are shown. The energy associated with these two simultaneous scatterings, $\hbar\omega \ll \Delta$. In the lower part creation of pairs of particles, both with momenta $|\Delta k| = k' - k$ are shown, these processes are important for $\hbar\omega \gtrsim 2\Delta$.

tion matrix element is the same for the two transitions. In the case of electromagnetic absorption (case II) the interaction is through the term $\vec{p}\cdot\vec{A}$ where $\vec{A}$ is the vector potential, and the reversal of the momentum $\vec{p}$ leads to the change of the sign of the interaction.

In order to see the processes which are involved let us look at transitions in which the scattering of quasiparticles is involved. For these processes terms like

$$[U_{k'}^* U_k + V_k^* V_{k'}]\gamma_{1,k'}^{\dagger}\gamma_{1,k} \tag{3.72a}$$

and

$$[U_{k'}^* U_k - V_{k'}^* V_k]\gamma_{1,k'}^{\dagger}\gamma_{1,k} \tag{3.72b}$$

arise. The first applies to ultrasonic attenuation, the second to electromagnetic absorption. From Eq. (3.71) the transition matrix element for the scattering of quasiparticles is then given by $(UU' \pm UV')$ and the creation or annihilation of pairs of quasiparticles by $(UU' \mp VU')$. The coherence factors are different from those appropriate for superconductors; the upper and lower signs apply for interactions which are either invariant or change sign with changing the momentum related to the interaction with the external applied potential.

The calculation of the transition matrix elements proceeds along the lines worked out for the superconducting ground state, and we obtain (Tinkham, 1975)

$$(UU' \pm VV')^2 = \tfrac{1}{2}\left(1 \pm \frac{\Delta^2}{EE'}\right) \tag{3.73}$$

for scattering of quasiparticles, while for the creation or destruction of pairs of quasiparticles

$$(VU' \mp UV') = \tfrac{1}{2}\left(1 \mp \frac{\Delta^2}{EE'}\right). \tag{3.74}$$

The transition rates between energy levels which differ by $\hbar\omega$ are given by

$$\frac{\alpha_{\text{CDW}}}{\alpha_N} = \frac{1}{\hbar\omega}\int_{-\infty}^{\infty} \frac{(E(E+\hbar\omega) \pm \Delta^2)[f(E) - f(E+\hbar\omega)]}{(E^2-\Delta^2)^{1/2}\left[(E+\hbar\omega)^2 - \Delta^2\right]^{1/2}} \tag{3.75}$$

where the + and − signs apply for case I and II coherence factors respectively.

The evaluation of this integral continues along the lines which have been worked out for the superconducting case (Tinkham, 1975 and references cited therein). In the low frequency limit, the electromagnetic absorption has a temperature dependence of

$$\frac{\alpha_{\text{CDW}}}{\alpha_N} = \frac{2}{1+\exp(\Delta/k_B T)} \tag{3.76}$$

and sharply decreases with decreasing temperature. The expected behavior is the same as that of the ultrasonic attenuation in superconductors. Ultrasonic attenuation, like electromagnetic absorption for superconductors, is expected to display a maximum somewhat below $T_{\text{CDW}}^{\text{MF}}$; the maximum depends on the excitation frequency and diverges at $T_{\text{CDW}}^{\text{MF}}$ as $\omega \to 0$.

In Eq. (3.70) the spin degrees of freedom are not included; for interactions, which involve spin flip processes, case II coherence factors occur. Consequently, we expect a maximum in the nuclear spin lattice relaxation rate somewhat below T_{CDW}^{MF}, in a fashion which is similar to the Hebel-Slichter peak in conventional superconductors.

3.3 Experimental Evidences for the Charge Density Wave Transition and Ground State

A broad variety of experiments have been conducted on the various inorganic linear chain materials which were discussed in Chapter 2 and they have been summarized in various reviews listed in the Appendix. The second order phase transition which leads to the charge density wave ground state has been examined by measurements of various thermodynamic quantities such as specific heat, thermal expansion, and elastic constants. The transition to a charge density wave ground state also leads to the development of a single particle gap at the Fermi level and is consequently accompanied by a metal-insulator transition. Therefore, the measurement of the various transport coefficients also gives clear evidence for the phase transition; the magnitude of the single particle gap can be extracted from these measurements. Finally, a variety of structural measurements, such as X-ray and neutron scattering and scanning tunneling microscope experiments, together with measurements of local properties (by nuclear magnetic resonance, for example), lead to the examination of the structural changes which accompany the phase transition. From these experiments the period and amplitude of the lattice distortion can also be extracted.

3.3.1 Transition Temperatures

Within the framework of mean field theory, the second order phase transition leads to a specific heat anomaly at T_{CDW}^{MF}, with a jump given by Eq. (3.60).

The specific heat measured in $K_{0.3}MoO_3$ over a broad temperature range is shown in Fig. 3.9 (Kwok et al., 1990). There is a well defined anomaly at $T = 183$ K and this identifies a phase transition to a three-dimensionally ordered ground state. This temperature, as will be discussed in Chapter 5, is different from the mean field transition temperature T_{CDW}^{MF} as given by Eq. (3.19) because of

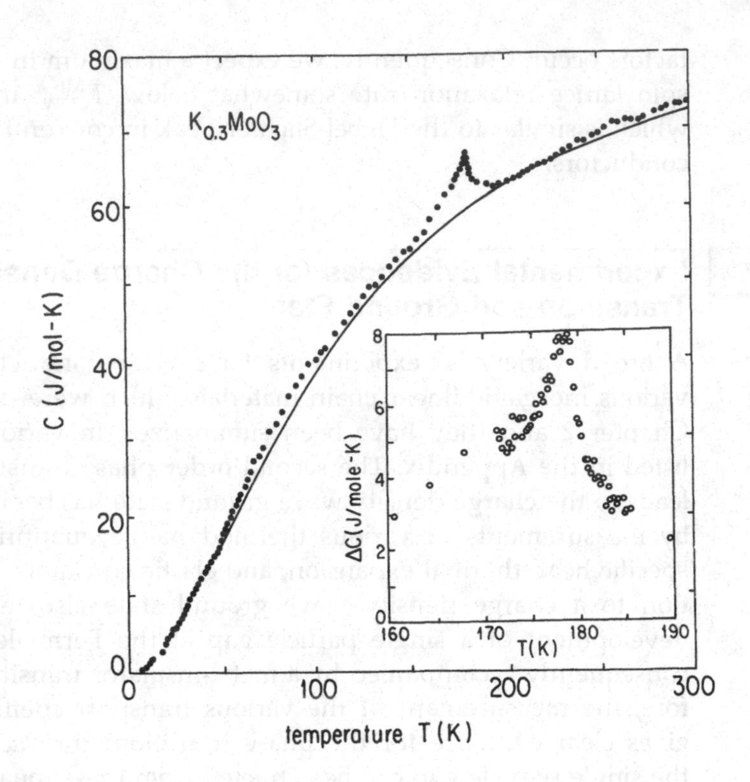

Figure 3.9. The temperature dependence of the specific heat of $K_{0.3}MoO_3$. (Kwok et al., 1990). The insert shows the specific heat near the transition, with a background subtracted.

fluctuation effects. Therefore, the transition temperature where a second order transition to the ground state with three-dimensional long range order occurs will be called T_{3D}. The peak which is observed at the transition is significantly larger than that predicted by Eq. (3.60); and instead of a sharp jump at T_{CDW} the transition is broad. Both features indicate the importance of fluctuation effects, as will be discussed in Chapter 5. Specific heat experiments have also been performed in other materials with a CDW ground state and in general the observations are similar to that shown in Fig. 3.9.

The metal-insulator transition can easily be detected by measurements of various transport coefficients. Figure 3.10 displays *dc* resistivity measurements conducted on various compounds (Grüner, 1988). In the metallic state, in the absence of a gap, the conductivity is expected to display a typical metallic behavior if

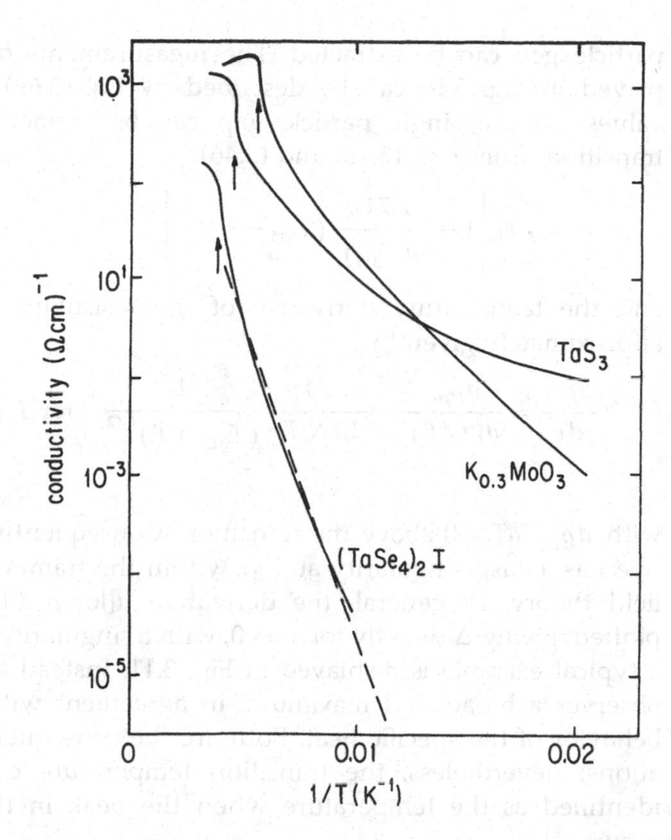

Figure 3.10. *dc* conductivity as a function of inverse temperature in some compounds with a charge density wave ground state (Grüner, 1988). The dotted line indicates the behavior given by Eq. (3.78).

no fluctuation effects are present. In the charge density wave state, the *dc* conductivity is described as

$$\sigma_{dc} = \mu n(T) \tag{3.77}$$

where μ is the mobility and $n(T) = n[1 - f_s(T)]$ with $f_s(T)$ given by Eq. (3.57). At low temperatures, where $|\Delta|$ is only weakly temperature dependent the *dc* conductivity has the following temperature dependence:

$$\sigma_{dc} = \mu(T)\exp\left(-\frac{\Delta}{k_B T}\right) \tag{3.78}$$

and with μ assumed to be temperature independent, the single particle gap can be extracted. The measurements of $\sigma_{dc}(T)$ displayed in Fig. 3.10 can be described by Eq. (3.69) from which values for the single particle gap can be extracted. Near the transition, from Eqs. (3.48) and (3.46),

$$\sigma_{dc} = \mu N_e \left[1 - \frac{1.74\pi}{4(T_{3D})^2} (T_{3D} - T)^{1/2} \right] \tag{3.79}$$

and the temperature derivative of the resistivity, $\rho_{dc} = \sigma_{dc}^{-1}$, is approximately given by

$$\frac{d\rho_{dc}}{dT} \simeq \frac{d\rho_{dc}}{d(1/T)} \simeq \frac{1.74\pi}{2\mu N_e T_c^2} \frac{1}{(T_{3D} - T)^{1/2}} \quad \text{for } T \lesssim T_{3D} \tag{3.80}$$

with $d\rho_{dc}/dT \simeq 0$ above the transition. Consequently, the derivative has a cusp singularity at T_{3D} within the framework of mean field theory. In general, the derivative $d[\log \rho_{dc}(T)]/d(1/T)$ is plotted, giving Δ directly for $T \to 0$, with a singularity at $T = T_{CDW}$; a typical example is displayed in Fig. 3.11. Instead of a cusp one observes a broadened maximum, in agreement with the overall behavior of the specific heat. Both are the consequence of fluctuations; nevertheless, the transition temperature can clearly be identified as the temperature when the peak in the derivative occurs.

The various transition temperatures and also the single particle gap values, as obtained from resistivity measurements for some compounds with a CDW ground state, are collected in Table 3.1. In all cases the value $2\Delta/k_B T_{3D}$ well exceeds the weak coupling BCS result of $2\Delta/k_B T_{3D} = 3.52$, suggesting that the transition temperatures are significantly lower than the mean field transition temperature, T_{CDW}^{MF}.

$NbSe_3$ is different from the materials which show metal-insulator transitions. Here the phase transition removes only part of the Fermi surface, and below T_{3D} the material remains a semimetal. Furthermore, the slightly different chains undergo phase transitions at different temperatures, called T_{3D}^1 and T_{3D}^2. The temperature dependence of the *dc* resistivity displayed in Fig. 3.12 (Monceau, 1987) clearly shows the two phase transitions.

Additional evidence for the development of a nonmagnetic, semiconducting ground state comes from magnetic susceptibility measurements. In the metallic state, $\chi(T)$ has the usual Pauli form

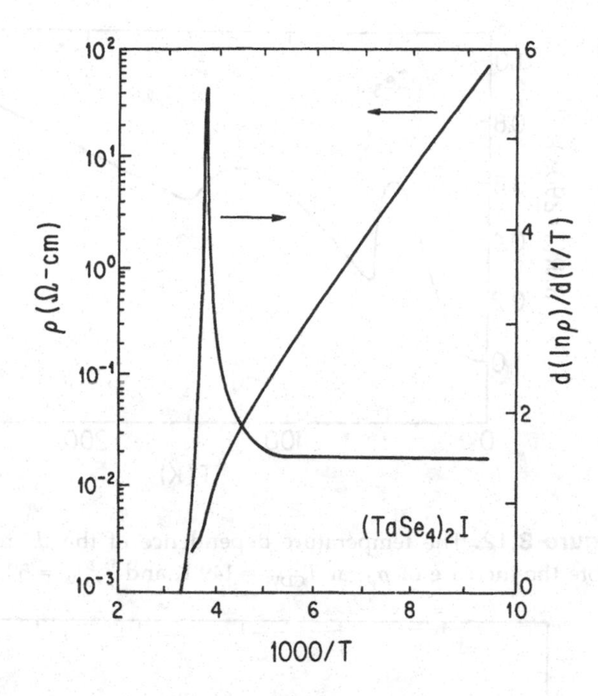

Figure 3.11. The temperature derivative of the *dc* resistivity in $(TaSe_4)_2I$ along the chain direction. The peak identifies the phase transition at $T =$ 263 K.

Table 3.1. Transition temperatures T_c, single particle gaps Δ_ρ (as obtained from the *dc* resistivity), Δ_{opt} (obtained from the optical conductivity, electron-phonon coupling constant λ (Eq. (3.44)), and coherence length ξ along the chain direction (Eq. 1.26)) for some linear chain compounds with a *CDW* ground state.

	$T_c(K)$	$2\Delta_\rho(K)$	$2\Delta_{opt}(K)$	λ	$\xi_\parallel(T=0)(\AA)$
KCP	189	1400	1700	0.302	118
$K_{0.3}MoO_3$	183	920	1400	0.34	125
TaS_3	215	1600	1800	0.65	40
$NbSe_3$	{ 145 59	700 [(a)]	n/a	0.53	61
$(TaSe_4)_2I$	263	3000	2900	0.43	100

(a) tunneling experiments

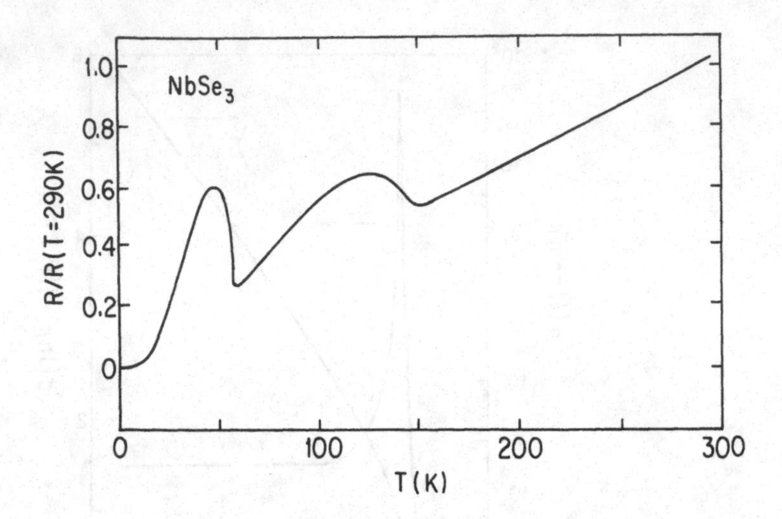

Figure 3.12. The temperature dependence of the *dc* resistivity of $NbSe_3$. Note the increase of ρ_{dc} at $T^1_{CDW} = 149$ K and $T^2_{CDW} = 59$ K.

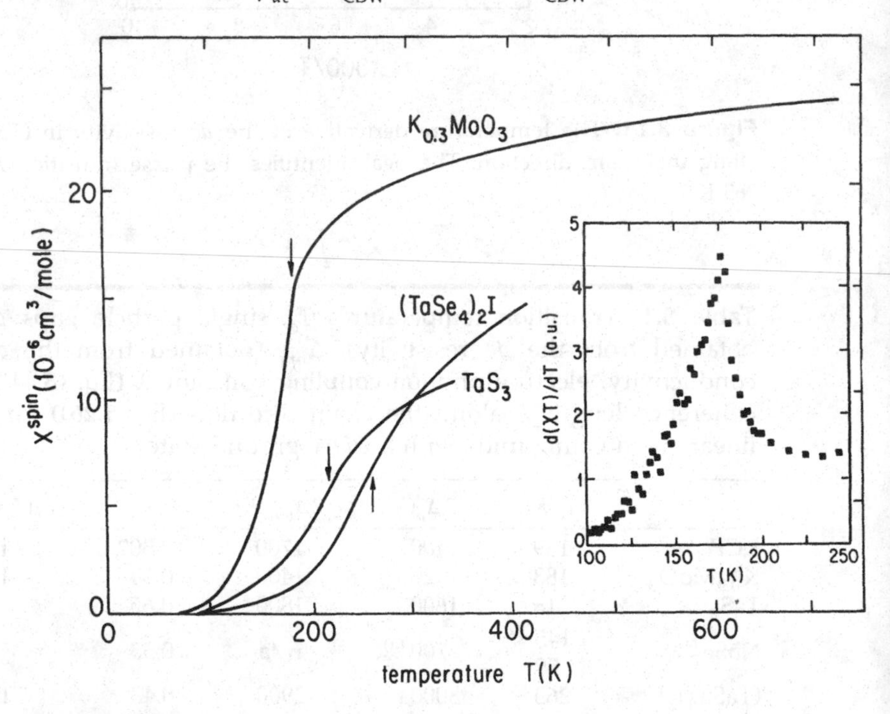

Figure 3.13. Temperature dependence of the magnetic susceptibility in several inorganic linear chain compounds with charge density wave ground states (Johnston et al., 1985). The arrows indicate the transition temperatures, and the insert shows the temperature derivative.

of Eq. (2.4), while below T_{3D} (which is proportional to the number of thermally excited carriers) the temperature dependence is given by Eq. (3.57). At low temperatures, $T \ll T_{3D}$, this reduces to

$$\chi(T) \simeq \frac{\mu_B^2}{T} \exp\left(-\frac{\Delta}{k_B T}\right) \tag{3.81}$$

where the weak temperature dependence of the gap well below T_{3D} has been neglected.

The temperature dependences of the magnetic susceptibility of various materials are displayed in Fig. 3.13 (Johnston et al., 1985). The susceptibility is somewhat temperature dependent even above the transition, which is indicated by the arrows for the various compounds. This is again due to fluctuations, and as for the electrical resistivity, the transition temperatures are identified by evaluating the derivative $d\chi(T)/d(1/T)$. Again, a rounded transition, such as that shown in the insert of Fig. 3.9, is observed.

3.3.2 The Kohn Anomaly

The phonon dispersion relation and its modification by electron-phonon interactions has been studied in various compounds by

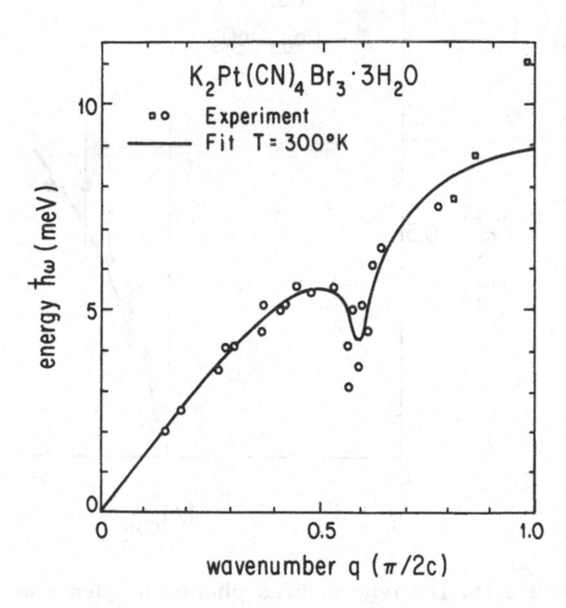

Figure 3.14. The phonon dispersion relation, measured above the charge density wave transition temperature in KCP (after Comes et al., 1975). For the details of the fit to the experimental results, see the original publication.

inelastic neutron scattering. The materials have a complicated crystal structure and consequently numerous phonon modes occur. Which of these interacts with the electron states depends on the electron-phonon coupling constants; the resulting distortions are difficult to calculate from first principles. In KCP the distortion involves a modulation of the repeat distances in the chain direction; in $K_{0.3}MoO_3$ it is related to the transverse displacement of some Mo atoms; and in $NbSe_3$ it is associated with an out-of-phase modulation of the different chains. In Fig. 3.14, the phonon spectrum is displayed for KCP at a temperature somewhat above T_{3D}. The sharp dip, at wavevector $q = 0.85\pi/a$ reflects the softening of the phonon branch and the full line is a fit to the renormalized phonon spectrum, including the renormalization of the phonon frequencies at wavevectors near $q = 2k_F$.

The temperature dependence of $\omega_{\mathrm{ren},2k_F}$, as given by Eq. (3.18), can be examined in detail by approximating the renormal-

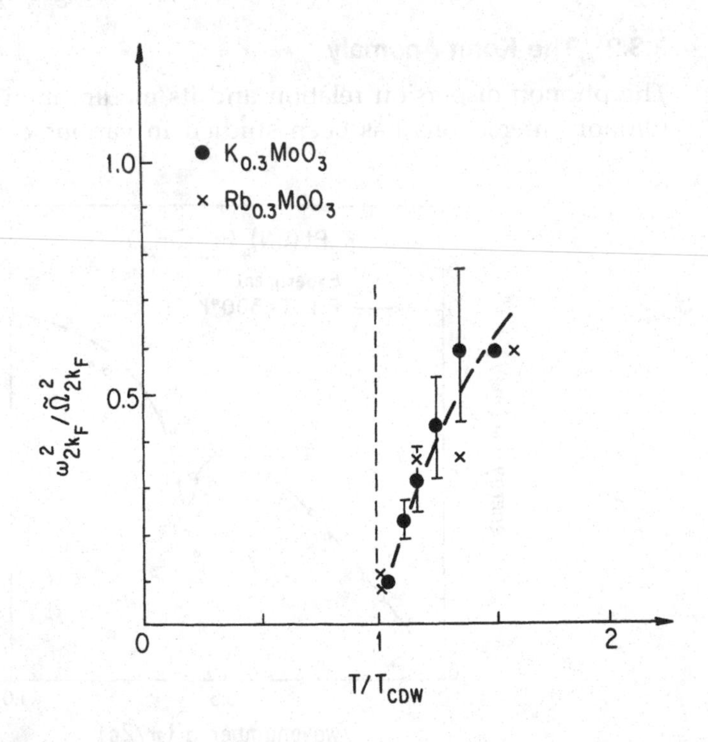

Figure 3.15. The renormalized phonon frequency as the function of temperature in $K_{0.3}MoO_3$. The dashed line is the prediction of the mean field theory, Eq. (3.21) (Pouget, 1987).

ized soft mode as a damped harmonic oscillator. Such an approach leads to a frequency dependent displacement-displacement correlation function, which for $q = 2k_F$ is given by (Pouget and Comes, 1989)

$$S(2k_F, \omega) = k_B T \frac{\Gamma_{2k_F}}{(\omega - \omega_{\text{ren},2k_F})^2 + 4\Gamma_{2k_F}^2} \tag{3.82}$$

where Γ_{2k_F} is the damping constant associated with the oscillations of the renormalized phonon frequency. The temperature dependence of $\omega_{\text{ren},2k_F}$ evaluated by fitting the measured scattering cross section to Eq. (3.82) is given in Fig. 3.15 for two compounds (Pouget, 1987). Close to T_{CDW}, $\omega_{\text{ren},2k_F}$ has a temperature dependence somewhat different from that predicted by mean field theory, Eq. (3.21). This suggests that fluctuations above the transition are important and will be discussed in Chapter 5.

3.3.3 The Energy Gap for $T < T_{\text{CDW}}$

The *dc* conductivity measured below the phase transition was found to be a semiconductor like in most of the materials and $\sigma_{dc}(T)$ can be adequately fitted to Eq. (3.79). Such fits, at temperatures well below the transition temperatures lead to single particle gaps, as seen through thermal excitations. The gap values obtained in this manner are collected in Table 3.1. The single particle gaps have also been studied by optical methods, which all clearly identify the gap structure. An example is shown in Fig. 3.16 where the frequency dependent conductivity evaluated both above and below $T_{3D} = 183$ K is shown. Above the transition the conductivity is Drude-like at low frequencies with an additional (most probably) interband transition. In contrast, at $T = 5$ K the conductivity is low below about 10^{-1} eV; indicating the appearance of the single particle gap. The gap, however, can only be evaluated by a detailed comparison of the measured frequency dependent conductivity and $\sigma(\omega)$ calculated for the Peierls state. Such a comparison will be made in Chapter 9. The gap values, quoted in Table 3.1, have been inferred from optical studies by performing such a comparison. In the material $NbSe_3$ optical studies cannot identify the gap structure because of the semimetallic character of the material for $T < T_{\text{CDW}}$, and here single particle tunneling (Ekino and Akimitsu, 1987; Fournel et al., 1986) has been used to evaluate the gap.

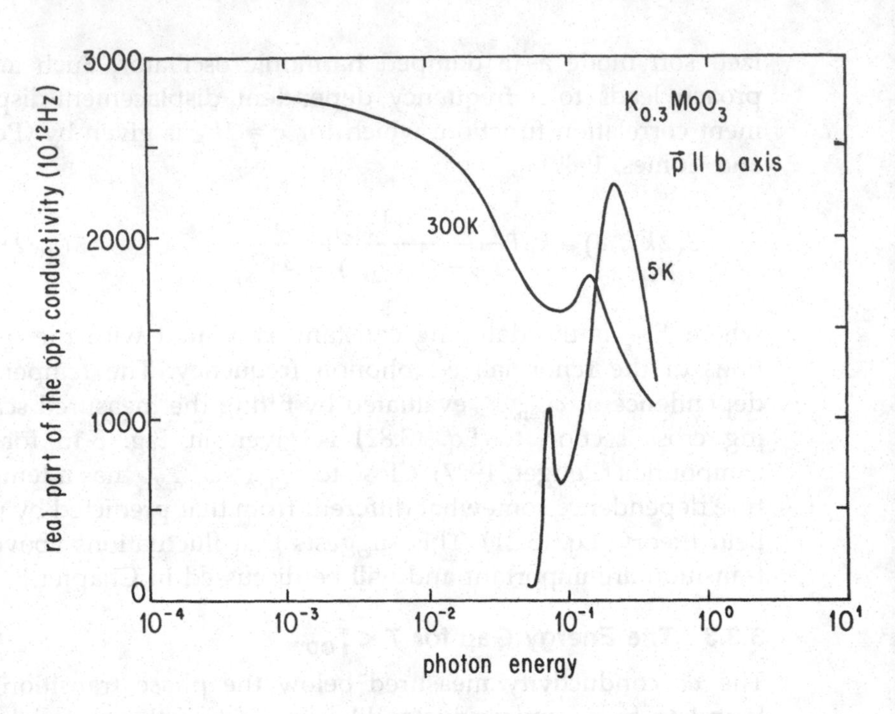

Figure 3.16. The optical conductivity of $K_{0.3}MoO_3$ measured along the chain direction both above and well below the charge density wave transition (Travaglini and Wachter, 1984; Degiorgi et al., 1991).

The measured single particle gaps, together with the Fermi energies evaluated in Chapter 2, lead by virtue of Eq. (3.44) to the electron-phonon coupling constant λ. In all cases λ is significantly smaller than one, indicating the weak coupling theory, for which $\Delta \ll \epsilon_F$, applies. Consequently, the likely explanation for the large single particle gaps (when compared with the transition temperature) is not strong coupling effects; but the fact that fluctuations due to the quasi-one-dimensional character of the materials strongly suppress the transition temperatures where the three-dimensional ordered ground state develops. Imperfect nesting, however, may also play a role (Huang and Maki, 1989).

The coherence lengths given by Eq. (1.26) can also be evaluated using the v_F values discussed in Chapter 2, together with the measured single particle gaps. The ξ_0 values obtained are also collected in Table 3.1. Because of the large gaps, the coherence length associated with the CDW state is short, extending only a few lattice constants. The coherence lengths shown in Table 3.1 are the values along the highly conducting direction. Since $\xi = \hbar v_F/\pi\Delta$, for an isotropic gap and strongly anisotropic band structure, the coherence lengths are significantly shorter in the directions perpendicular to the chains. The band structure anisotropy for $K_{0.3}MoO_3$, as discussed in Chapter 2, leads to coherence lengths $\xi_\perp(2a-c) = 7$ Å and $\xi_\perp(2a+c) = 1.5$ Å in the two perpendicular directions. This substantial anisotropy also leads to a rather small coherence volume

$$[\xi_\| \cdot \xi_\perp(2a-c) \cdot \xi_\perp(2a+c)] \sim 420 \text{ Å}^3 \tag{3.83}$$

indicating that this volume contains only a few electron-hole pairs. Similar anisotropies have been found for other materials which develop a charge density wave ground state. Consequently the coherence lengths perpendicular to the highly conducting direction are small in these compounds as well.

3.3.4 The Lattice Distortion for $T < T_{CDW}$

Below the transition a static modulation of the ionic positions occurs and, for a linear chain of atoms, the positions are given in the case of long range order by

$$X_n = na + \Delta u(T)\sin(n2k_F a + \phi) \tag{3.84}$$

where the amplitude of the distortion $\Delta u(T)$ is determined by Eq. (3.23); it has the same temperature dependence as the order parameter Δ. The new periodicity at $q = 2k_F$ leads to new Bragg peaks, the positions of which give k_F directly. The intensities are related to the displacement Δu (Eagen et al., 1975).

In the absence of distortion, the scattering intensity is given by

$$I(k) = e^{-2W}\sum_H \delta(k - Ha^*) \tag{3.85}$$

where $a^* = 2\pi/a$, e^{-2W} is the Debye-Waller factor, and H is the one-dimensional Miller index of reflection. Using Eq. (3.84) we can calculate the nth order static structure factor $S_n(Ha^*)$, and the scattering intensity

$$I \simeq \sum_{n,H} S_n(H)\delta(k + n2k_F - Ha^*). \tag{3.86}$$

For a small amplitude lattice distortion the first term gives

$$I^0(k) = e^{-2W} \sum_H \delta(k - Ha^*)\left[1 - \tfrac{1}{2}Ak^2\right] \tag{3.87}$$

with A constant, describing the scattering intensity of the Bragg points. For small Δu the second term in the bracket is small, and the intensity of the Bragg reflections is only slightly modified by the formation of the periodic lattice distortion. The next term in Eq. (3.86) gives

$$I^{\pm}(k) = e^{-2W} \sum_H \delta(k - Ha^* \pm 2k_F)\tfrac{1}{2}(k\Delta u^2). \tag{3.88}$$

This describes a set of lattice points with a wavevector given by $k \pm 2k_F$ and with an intensity proportional to Δu^2, and thus proportional to the square of the order parameter. Therefore, the positions of the new peaks, which arise as a consequence of the new periodicity, lead to the period of the lattice modulation and, though the relation $\lambda_0 = \hbar / k_F$, to the Fermi wavevector. The measured intensity of these peaks allows for the evaluation of the temperature dependence of the order parameter (Pouget and Comes, 1989 and references cited therein).

For a three-dimensional order which develops in a system of (weakly) coupled chains the period of the lattice distortion is given by $\vec{Q}_{CDW} = (2k_F, Q_\perp)$; where $Q_\perp$ describes the ordering of the charge density wave modulation on the neighboring chains. For a system of coupled chains the expression analogous to Eq. (3.88) reads

$$I^{\pm}(k) = e^{-2W} \sum \delta\left(\vec{k} - \vec{R}^* \pm \vec{Q}_{CDW}\right)\tfrac{1}{2}(k\Delta u^2) \tag{3.89}$$

where $\vec{R}^*$ is a reciprocal lattice vector. Consequently, the new superlattice reflection points give $2k_F$ and $Q_\perp$ respectively.

Both X-ray and elastic neutron scattering have been used to evaluate the wavevector associated with the lattice distortion and the magnitude of the distortion as the function of temperature (these have been summarized by Pouget and Comes, 1989; and Pouget, 1987). The wavevectors associated with the ground state are collected in Table 3.2, where the first column always refers to the chain direction and the next two columns to directions perpendicular to the chains. In all cases the period λ_0 is incommensurate with the underlying lattice. Also, in the majority of cases

Table 3.2. Structural distortion parameters for the *CDW* state of various compounds. See P. Monceau (1985) and references cited therein.

	Superstructure			$\Delta u_{max}(\mathring{A})$ *(exp)*	$\Delta u_{max}(\mathring{A})$ *(Eq. 3.90)*
	$q_\parallel$	$q_\perp$			
KCP	0.3	0.5	0.5	0.027	0.049
$K_{0.3}MoO_3$	0.71	0.5	0.5	0.05	0.035
TaS_3	~ 0.25			n/a	—
$NbSe_3$	0.24	0	0	n/a	—
	0.26	0.5	0.5		
$(TaSe_4)_2I$	0.084	0.05	0.05	0.09	

the CDW modulations on the neighboring chains are antiphase. That is, the maxima of the CDW modulations on one chain are accompanied by minima on the neighboring chains. The more complicated structural arrangements such as those which occur in $(TaSe_4)_2I$ have not been accounted for to date. The temperature dependence of the lattice distortion is shown for some compounds in Fig. 3.17. In all cases a temperature dependence reminiscent of the temperature dependence of the order parameter as given by Eq. (3.19) is obtained. The full line on the figure is $|\Delta(T)|$ as given by the weak coupling BCS theory. The displacement amplitudes, also obtained from the scattering intensity, are summarized in Table 3.2. The amplitude is given in terms of the electron-phonon coupling constant, the unrenormalized phonon frequency, and the density of states as

$$\Delta u = \frac{2\Delta}{g}\left(\frac{\hbar}{2NM\omega_{2k_F}}\right)^{1/2}. \qquad (3.90)$$

The parameters which appear on the right hand side have been evaluated before and are shown in Tables 2.1 and 3.1. The unrenormalized phonon frequency has been obtained from neutron scattering experiments and M has been chosen to be the mass of the basic structural unit (see Table 2.1). Consequently, Δu can be evaluated and the values calculated for the materials on which the amplitude of the lattice modulation was measured are also collected in Table 3.2.

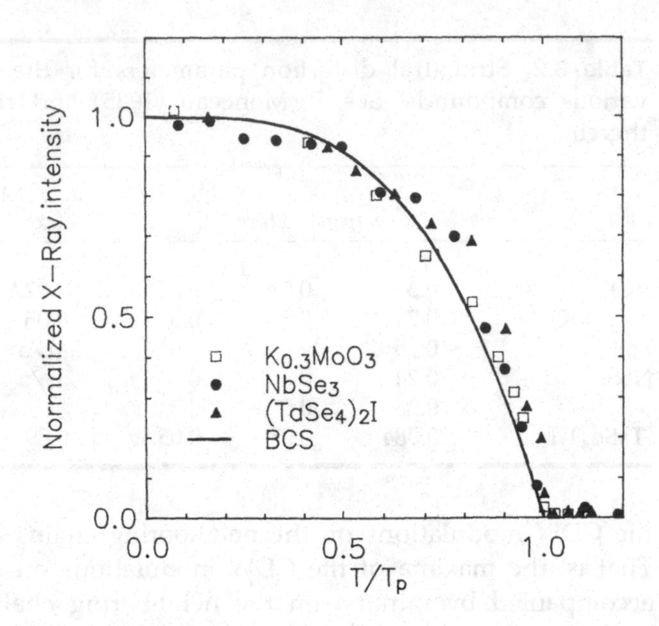

Figure 3.17. Temperature dependence of the lattice distortions in $K_{0.3}MoO_3$, $NbSe_3$, and $(TaSe_4)_2I$. The full line is $\Delta(T)$ given by the weak coupling BCS limit (Grüner, 1988 and references given therein).

The periodic modulation of the charge density leads to a periodic modulation of electric field gradient at the nuclear sites; and consequently to changes in the nuclear magnetic resonance line shape. For the charge density wave as given by Eq. (3.52), to first approximation, the resonance frequency at the nuclear site $\vec{R}$ is

$$\nu(\vec{R}) = \nu_0 + \delta\nu(R) = \nu_0 + \nu_1 \cos\left(\vec{Q}_{\mathrm{CDW}} \cdot \vec{R} + \phi\right) \qquad (3.91)$$

where ν_0 is the resonance frequency in the absence of the charge density wave and ν_1 is proportional to the amplitude of the density wave. The resonance line shape which results from the distribution of the resonance frequencies can readily be calculated (Berthier and Segransan, 1987 and references cited therein) and gives

$$g(\nu) = \mathrm{const} \int_{\nu = \mathrm{const}} \frac{ds}{\vec{\nabla}\left(\delta\nu(\vec{R})\right)}. \qquad (3.92)$$

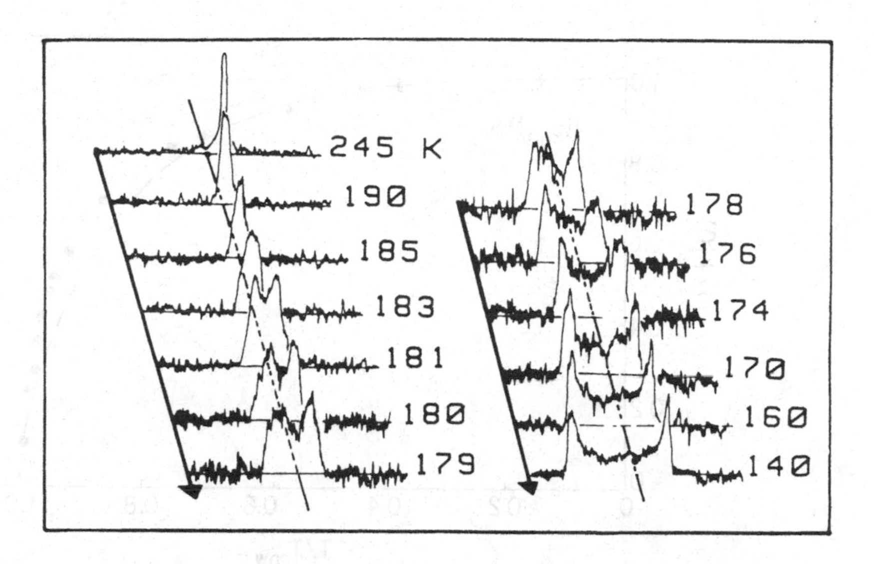

Figure 3.18. Nuclear magnetic resonance line shapes in $Rb_{0.3}MoO_3$ at several temperatures. The distance between the "wings" of the distribution is proportional to the amplitude of the density modulation (Berthier and Segransan, 1987).

For an incommensurate CDW we obtain (Ross et al., 1986; Janossy et al., 1987)

$$g(\nu) = \text{const} \cdot \nu_1 \left[1 - \left(\frac{\nu}{\nu_1} \right)^2 \right]^{-1/2} \qquad |\nu| < \nu_1 \qquad (3.93)$$

$$g(\nu) = 0 \qquad |\nu| > \nu_1 \qquad (3.94)$$

Typical experiments on $Rb_{0.3}MoO_3$ are shown in Fig. 3.18. Above the transition the linewidth is determined by interactions different from the electric field gradient induced distribution, and below T_{3D}, $g(\nu)$ can be described by Eq. (3.93). The analysis leads to the temperature dependence of the amplitude of the density modulation; i.e., the temperature dependence of the order parameter (shown in Fig. 3.19) with the full line is the temperature dependence of the order parameter as given by the weak coupling BCS expression. The latter accounts for the experimental results except close to the transition temperature; where a critical exponent different from $\eta = \frac{1}{2}$ as expected for a mean field

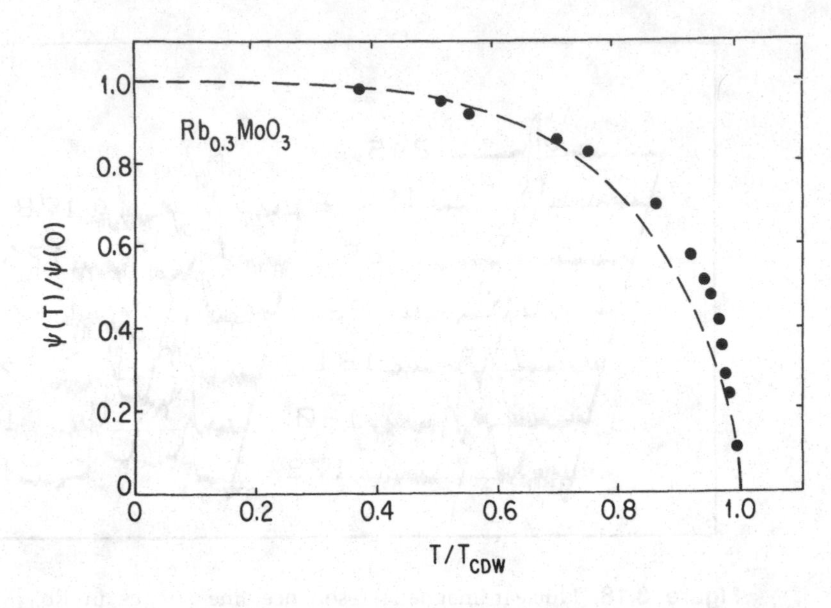

Figure 3.19. Temperature dependence of the amplitude of the density wave modulation in $Rb_{0.3}MoO_3$ (full circles). The dashed line is the temperature dependence of the order parameter as given by the weak coupling BCS theory (Berthier et al., 1992).

transition (see Eq. (3.62a)) is observed. This is most probably due to fluctuation effects.

Charge density waves have also been observed by scanning tunneling microscopy in various materials (Slough et al., 1989; Dai et al., 1991 and references cited therein). These measurements also give information about the period of the density wave modulation. The magnitude of the modulation, however, cannot be extracted from these measurements.

The Spin Density Wave Transition and Ground State: Mean Field Theory and Some Basic Observations

...that something is structured is not the essential thing here, but rather why and how it is structured.

—Václav Havel *Letters To Olga*

4

The spin density wave ground state of metals is thought to arise as a consequence of electron-electron interactions. Questions about the consequences of this interaction on the Fermi-liquid state have received considerable interest. There is also a great deal of literature on the various ground states which develop, and on the nature of low lying excitations, which occur in one-dimensional models (see for example, Anderson, 1990). The simplest possible description of this interaction is given by the following term in the Hamiltonian

$$\mathscr{H}_{\text{int}} = \frac{U}{N} \sum_{q,\sigma} n_{q,\sigma} n_{-q,-\sigma} = \frac{U}{N} \sum_{k,k',q} a^{\dagger}_{k,\sigma} a_{k+q,\sigma} a^{\dagger}_{k',-\sigma} a_{k'-q,-\sigma} \tag{4.1}$$

where U is the on-site Coulomb interaction. Together with the kinetic energy term, the so-called Hubbard Hamiltonian is given by

$$\mathscr{H} = \sum_{k,\sigma} \epsilon_k a^{\dagger}_{k,\sigma} a_{k,\sigma} + \frac{U}{N} \sum_{k,k',q} a^{\dagger}_{k,\sigma} a_{k+q,\sigma} a^{\dagger}_{k',-\sigma} a_{k'-q,-\sigma}. \tag{4.2}$$

Many of the important aspects of the one-dimensional Hubbard model (see also Lieb and Wu, 1968), such as spin-charge separation, etc., will not be discussed here; we will simply assume that the 1D electron gas can be described in terms of a conventional Fermi liquid with the response functions as discussed in Chapter 1, and confine ourselves to the mean field solution. As mentioned in the previous chapter this is done by arguing that while fluctuations are important, nevertheless for a system of coupled chains the ground state is similar to that given by the mean field solution. (Whether this argument is correct or not remains to be seen; the experimental results discussed later are in fair agreement with those obtained from a mean field treatment.) As first discussed by Overhauser (1960, 1962) the ground state of Eq. (4.2) is a spin density wave state, and the transition together with the parameters of the state will be described within the framework of a mean field theory, with the assumption that $U << \epsilon_F$; which is equivalent to the weak coupling limit for superconductors. The ground state, as is the case for the CDW state, opens up a gap at the Fermi level in both spin subbands and, consequently, in the case of the complete removal of the Fermi surface (such as expected for a highly anisotropic band structure), a metal insulator transition results. The ground state also has a well defined magnetic response, similar to that of an antiferromagnet. Consequently, transport and magnetic measurements, together with local probes which sample the internal magnetic fields at the nuclear sites, have been used to evaluate the essential characteristics of the ground state.

4.1 Mean Field Theory of the Spin Density Wave Transition

The interaction between the electrons with opposite spin, as given by Eq. (4.1), leads to an enhanced response of the electron gas to an external magnetic field. This response can be simply described by using the mean field approximation. Assume that we apply an external magnetic field which varies along the chain direction as

$$H(x) = \sum_q H_q e^{iqx}. \tag{4.3}$$

The coupling to this field is described by an extra term in the

Hamiltonian

$$\mathcal{H}' = \sum_q M_q H_{-q} \tag{4.4}$$

where M_q is the qth component of the magnetization. We also assume that the magnetic field is applied to the (arbitrary) z direction. The spin direction parallel (opposite) to H is denoted by $\uparrow$ ($\downarrow$). Then the expectation value of the magnetization is

$$\langle M_q \rangle = \mu_B(\langle n_{q,\uparrow}\rangle - \langle n_{q,\downarrow}\rangle) = \chi_0(q) H_q^{\text{eff}} \tag{4.5}$$

with

$$H_q^{\text{eff}} = H_q + \frac{U(\langle n_{q,\uparrow}\rangle - \langle n_{q,\downarrow}\rangle)}{2\mu_B} \tag{4.6}$$

where $\chi_0(q)$ is the wavevector dependent susceptibility in the absence of Coulomb interactions. The self-consistent equation for the difference $\Delta n_q = \langle n_{q,\uparrow}\rangle - \langle n_{q,\downarrow}\rangle$ is then

$$\mu_B \Delta n_q = \chi_0(q)\left(H_q + \frac{U\Delta n_q}{2\mu_B}\right) \tag{4.7}$$

The magnetization from Eqs. (4.5) and (4.7) is

$$\langle M_q \rangle = \frac{\chi_0(q)}{1 - \dfrac{U\chi_0(q)}{2\mu_B^2}} H_q = \chi_{\|}(q) H_q \tag{4.8}$$

and the response is described as that of an electron gas with an enhanced susceptibility

$$\chi_{\|}(q) = \frac{\chi_0(q)}{1 - \dfrac{U\chi_0(q)}{2\mu_B^2}} \tag{4.9}$$

For a uniform magnetization, $q = 0$ and with $\chi_0(0) = 2\mu_B^2 n(\epsilon_F)$, we obtain a static susceptibility

$$\chi(0) = \frac{2\mu_B^2 n(\epsilon_F)}{1 - Un(\epsilon_F)}, \tag{4.10}$$

which is enhanced by the well known Stoner factor. For a one-dimensional electron gas $\chi_0(q)$ is strongly peaked at $q = 2k_F$, and the enhancement is most important for perturbations with this wavevector. $\chi_0(2k_F, T)$ is strongly temperature-dependent (see Eq. (1.17)), which leads to a strongly temperature dependent response, and (with fluctuation effects neglected) to a phase transition at temperature $T_{\mathrm{SDW}}^{\mathrm{MF}}$ defined by

$$\frac{U\chi_0(2k_F,T)}{2\mu_B^2} = Un(\epsilon_F)\ln\frac{1.14\epsilon_0}{k_B T} = 1. \tag{4.11a}$$

This gives

$$k_B T_{\mathrm{SDW}}^{\mathrm{MF}} = 1.14\epsilon_0 \exp(-1/\lambda_e), \tag{4.11b}$$

where the dimensionless electron-electron coupling constant is defined as

$$\lambda_e = Un(\epsilon_F). \tag{4.12}$$

Below $T_{\mathrm{SDW}}^{\mathrm{MF}}$ a static, spatially varying magnetization develops. By introducing the spatially dependent operators

$$\Psi_\sigma(x) = \sum_k e^{ikx} a_{k,\sigma} \tag{4.13}$$

the spin density is given as

$$S(x) = \frac{1}{2}\left[\Psi_\uparrow^\dagger(x)\Psi_\uparrow(x) - \Psi_\downarrow^\dagger(x)\Psi_\downarrow(x)\right] \tag{4.14}$$

$$= \frac{1}{2}\sum_{k,k'}\left\{a_{k,\uparrow}^\dagger a_{k',\uparrow} - a_{k,\downarrow}^\dagger a_{k',\downarrow}\right\} e^{-i(k-k')x} \tag{4.15}$$

Because of the divergent response function at $q = 2k_F$, we assume that only terms with $k' = k \pm 2k_F$ are important. Thus the above equation yields the expectation value

$$\langle S(x)\rangle = \frac{1}{2}\sum_k \left\{\langle a_{k,\uparrow}^\dagger a_{k+2k_F,\uparrow}\rangle - \langle a_{k,\downarrow}^\dagger a_{k+2k_F,\downarrow}\rangle\right\} e^{i2k_F x} + c.c. \tag{4.16}$$

We write the expectation value as

$$S = |S|e^{i\phi} = \langle n_{2k_F,\uparrow}\rangle = -\sum_k \langle a_{k,\uparrow}^\dagger a_{k+2k_F,\uparrow}\rangle = -\sum_k \langle a_{k,\downarrow}^\dagger a_{k+2k_F,\downarrow}\rangle = -\langle n_{2k_F,\downarrow}\rangle \tag{4.17}$$

assuming that the density modulations have the opposite sign

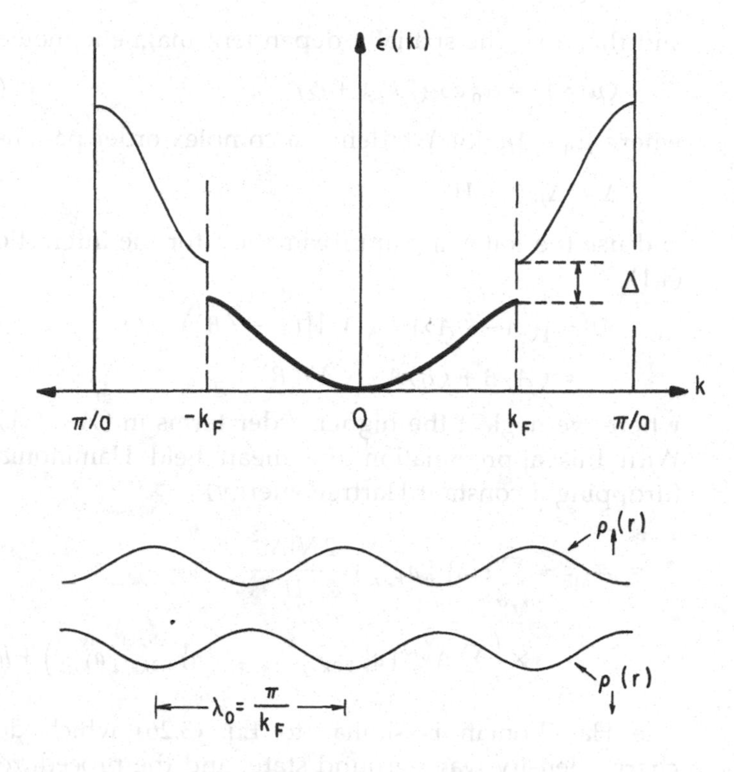

Figure 4.1. The dispersion relation and charge and spin density wave modulation for the two spin subbands in the spin density wave ground state.

(i.e., the same phase ϕ) in the two subbands. The situation, where the phases of the density modulations are different, leads to a complex spin density wave structure (such as spiral spin density waves); such states will not be discussed here. From Eq. (4.16) we obtain

$$\langle S(x)\rangle = 2|S|\cos(2k_F x + \phi). \tag{4.18}$$

Therefore, the spin density is spatially varying, with a period $\lambda_0 = \pi/k_F$ determined by the Fermi wavevector. There is no charge density modulation as $\langle n_{2k_F\uparrow}\rangle + \langle n_{2k_F\downarrow}\rangle$ does not show any spatial variation; the situation is displayed in Fig. 4.1.

The magnetic moment μ is given as

$$\langle \mu\rangle = g\mu_B\langle S\rangle = 2\mu_B\langle S\rangle \tag{4.19a}$$

and therefore the spatially dependent magnetic moment is

$$\langle \mu(x) \rangle = \mu_0 \cos(2k_F x + \phi) \tag{4.19b}$$

where $\mu_0 = 4\mu_B|S|$. We define a complex order parameter

$$\Delta = |\Delta| e^{i\phi} = US \tag{4.20}$$

and use the following approximation for the interaction from Eq. (4.1).

$$\begin{aligned} AB &= [(A - \langle A \rangle) + \langle A \rangle][(B - \langle B \rangle) + \langle B \rangle] \\ &\approx \langle A \rangle B + \langle B \rangle A - \langle A \rangle \langle B \rangle \end{aligned} \tag{4.21}$$

where we neglect the higher order terms in $(A - \langle A \rangle)(B - \langle B \rangle)$. With this approximation the mean field Hamiltonian becomes (dropping a constant Hartree energy)

$$\begin{aligned} \mathscr{H}_{\mathrm{MF}} = \sum_{k,\sigma} \epsilon_k a^{\dagger}_{k,\sigma} a_{k,\sigma} + \frac{2N|\Delta|^2}{U} \\ \times \left\{ \sum_k \Delta e^{i\phi} \left(a^{\dagger}_{k+2k_F,\uparrow} a_{k,\uparrow} + a^{\dagger}_{k+2k_F,\downarrow} a_{k,\downarrow} \right) + h.c. \right\}. \end{aligned} \tag{4.22}$$

The Hamiltonian is similar to Eq. (3.26) which describes the charge density wave ground state, and the procedure for obtaining the single particle gap, discussed in the previous chapter, can be adopted here. Straightforward diagonalization similar to that done for charge density waves leads to

$$\mathscr{H} = \sum_{k,\sigma} E_k \gamma^{\dagger}_{k,\sigma} \gamma_{k,\sigma} + \frac{2|\Delta|^2 N}{U} \tag{4.23}$$

with the dispersion relation for the operators $\gamma_{k,\sigma}$ given by

$$E_k = \epsilon_F + \operatorname{sign}(k - k_F)\left[\hbar^2 v_F^2 (k - k_F)^2 + \Delta^2\right]^{1/2} \tag{4.24}$$

for both spin subbands.

The development of the single particle gap leads to a decrease of the kinetic energy, while the interaction term $E_M = 2|\Delta|^2/U$ (the second term in Eq. (4.22)) gives a positive energy for a finite magnetization. The equilibrium gap value is obtained from

$$\delta E_{\mathrm{el}} + \delta E_M = 0 \tag{4.25}$$

and straightforward algebra similar to that which led to Eq. (3.44) gives

$$\Delta(T=0) = 2\epsilon_F \exp(-1/\gamma_e). \tag{4.26}$$

A comparison with Eq. (4.11b) leads to the weak coupling BCS relation $2\Delta = 3.52 k_B T_{SDW}^{MF}$ and as for charge density waves, the ground state energy is given by

$$E = -\frac{1}{2} n(\epsilon_F)|\Delta|^2. \tag{4.27}$$

The temperature dependence of the single particle gap, and also the amplitude of the spin density wave is given by the weak coupling BCS expression, and $\Delta(T)$ together with the temperature dependent condensate density are given by Eqs. (3.55) and (3.57).

Crudely speaking, the SDW ground state can be viewed as two charge density wave states—one for the "spin up" and one for the "spin down" subbands as shown in Fig. 4.1, with the charge density wave modulations given by

$$\rho_\uparrow(x) = \rho_0\left[1 + \frac{\Delta}{v_F k_F \gamma_e}\cos(2k_F x + \phi)\right] \tag{4.28a}$$

$$\rho_\downarrow(x) = \rho_0\left[1 + \frac{\Delta}{v_F k_F \lambda_e}\cos(2k_F x + \phi + \pi)\right]. \tag{4.28b}$$

The resulting spin density variation $\rho_\uparrow(x) - \rho_\downarrow(x)$, given by Eq. (4.18) with $|S| = |\Delta| N/U$ (see Eq. (4.21)), and the resulting charge density wave variation $\rho_\uparrow(x) - \rho_\downarrow(x) = \rho_0 =$ const are shown in Fig. 4.1. Both subbands however are tied to the Fermi surface; which will have important implications for the excitations and for the dynamics of the SDW ground state.

The above considerations are for an arbitrary spin orientation. In the case of a continuous spin symmetry the condensate density can be written as

$$\langle \vec{S}(x) \rangle = 2\vec{S}\cos(2k_F x + \phi) \tag{4.29}$$

where $\vec{S} = (S_x, S_y, S_z)$. Consequently, both the spin rotational and the translational symmetries are broken in the spin density wave state. This will have important consequences on the magnetic properties of the ground state and on the dynamics of the collective mode.

The description of the transition and ground state as given above is based on a strictly one-dimensional model. The materials which develop an SDW ground state, however, have significant transfer integrals perpendicular to the chain directions, and models for which this so called quasi-one-dimensional character is included are expected to be more appropriate. Taking into account the finite bandwidth in the perpendicular directions, the dispersion relation is given by (Yamaji, 1982, 1983; Huang and Maki, 1990).

$$\epsilon_k = -2t_a \cos ak_1 - 2t_b \cos bk_2 - 2t_c \cos ck_3 - \mu \tag{4.30}$$

where μ is the chemical potential and t is the transfer integral in the various directions in the tight binding approximation. For $t_a >> t_b, t_c$ the condition for nesting

$$\vec{\epsilon}_k = -\vec{\epsilon}_{k+Q} \tag{4.31}$$

with $\vec{Q} = (2k_F, \pi/b, \pi/c)$ is nearly, but not completely satisfied, and this leads to deviations from the various expressions which have been obtained on the basis of strictly one-dimensional models. The transition temperature $T^{\mathrm{MF}}_{\mathrm{SDW}}$ decreases with increasing transfer integrals perpendicular to the chain direction, and the transition is completely removed for

$$\Delta(T=0) = \frac{t_b^2}{2} \cos(ak_F/t_a)\sin^2(ak_F) \tag{4.32}$$

with $\Delta(T=0)$ given by Eq. (4.26). Also, the spin density wave modulation is given by

$$\vec{S}(\vec{r}) = 2\vec{S}\cos\left(\vec{Q}\cdot\vec{r} + \phi\right) \tag{4.33}$$

with $\vec{Q} = (2k_F, \pi/b, \pi/c)$.

The behavior of the order parameter and of the various thermodynamic quantities can be described near $T^{\mathrm{MF}}_{\mathrm{SDW}}$ in terms of the Ginzburg-Landau theory; in a fashion which follows the description of the charge density wave transition near $T^{\mathrm{MF}}_{\mathrm{CDW}}$. Consequently, $\Delta(T)$, and ΔC are also given by Eqs. (3.62) and (3.63) in terms of the coefficients a and b of the Ginzburg-Landau free energy functional.

The evaluation of coherence factors proceeds along the same lines used for the charge density wave ground state, and the resulting transition probabilities, agree with those calculated for charge density waves. The ultrasonic attenuation reflects the case II coherence factors, and is expected to display a maximum below T_{SDW}^{MF}; while the conductivity is determined by case I coherence, and is expected to drop sharply below the transition.

4.2 Experimental Evidences for the Spin Density Wave Transition and Ground State

Evidence of transitions to a spin density wave state have been found in several of the organic linear chain compounds which were discussed in Chapter 2. Measurements of the various thermodynamic quantities have been performed only in a few cases and the principal observations on the transition have been in the transport and magnetic properties.

As for charge density waves, the development of the SDW ground state opens up a gap at the Fermi level, leading to a metal-insulator transition. The *dc* resistivity measured on three compounds which develop SDW ground states is displayed in Fig. 4.2. In all three cases there is a well defined transition from metallic to semiconducting behavior; and as for charge density waves, the transition temperatures, called T_{3D}, are identified through the measurement of the temperature derivative $d \ln \rho / d(1/T)$ (see Eq. (3.80)). Below T_{3D}, the conductivities are well described by Eq. (3.78), and the single particle gaps, together with the transition temperatures are collected in Table 4.1. These values, together with the parameters which characterize the metallic state above T_{3D}, lead, using Eq. (1.26), to ξ_0 and by virtue of Eq. (4.12), to the coupling constant λ_e.

Values of the density of states above the transition, $n(\epsilon_F)$, have been evaluated from the magnetic susceptibilities measured above T_{3D}; and the values as given in Table 2.2 lead, using Eqs. (4.11) and (4.12), to the Coulomb energies (displayed in Table 4.1), which are significantly smaller the ϵ_F. In all of these compounds therefore, Coulomb effects lead to an SDW state, in the so-called weak coupling limit. The measured single particle gaps, together with the Fermi velocities, also lead to large coherence lengths, and the zero temperature values $\xi_{\parallel} = \hbar v_F / \pi \Delta$ are given in Table 4.1. As for charge density waves, because of the strongly

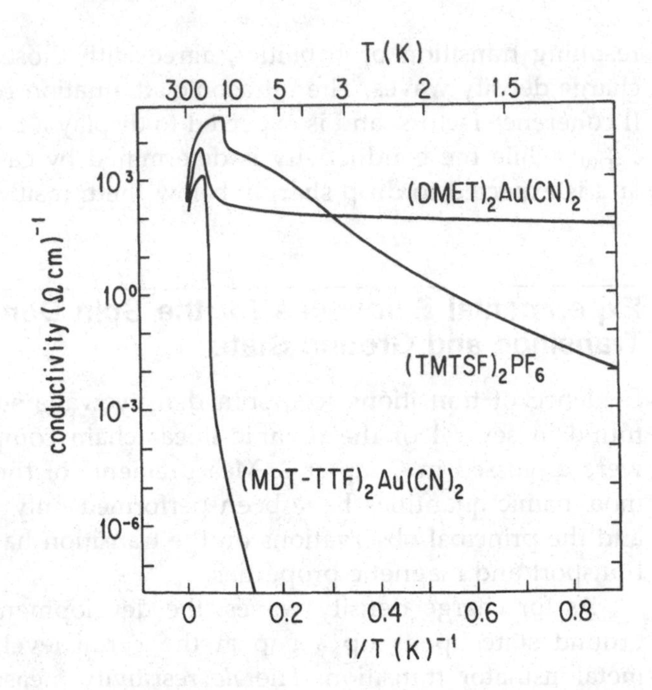

Figure 4.2. The temperature dependence of the *dc* conductivity in various materials which develop an SDW ground state at low temperatures. The TMTSF and MDT–TTF salts are semiconducting below $T = 11.5$ K and 20 K respectively, the DMET salt remains semimetallic below the transition (Bechgaard, et al., 1980; Kikuchi et al., 1987; Nakamura et al., 1990).

anisotropic bandwidth in the various crystallographic directions, the coherence length is also anisotropic. This anisotropy has been measured only for the $(TMTSF)_2X$ salts, and with the anisotropic Fermi velocities, as discussed in Chapter 2 we obtain $\xi_{\perp y} \simeq 30$ Å, while in the z direction $\xi_{\perp z} < c$ is the lattice constant.

The above values of the single particle gaps and Coulomb energies also suggest small magnetic moments associated with the spin density wave state. The amplitude of the SDW modulation may be written as

$$\frac{\mu}{\mu_B} = \frac{4|\Delta|}{U}. \tag{4.34}$$

The values of Δ and U as evaluated above lead to the reduced

Table 4.1. Parameters of the Spin Density Wave State in various compounds. Δ_ρ refers to the gap as evaluated from the temperature dependence of the *dc* resistivity.

	$T_{SDW}(K)$	$2\Delta_\rho(K)$	λ Eq. (4.11)	$\xi_\parallel(\mathring{A})$	μ/μ_b (NMR)
$(TMTSF)_2PF_6$	11.5[1]	30	0.26	320	0.08[4,5]
$(MDT\text{-}TTF)_2Au(CN)_2$	20[2]	NA			0.1[6]
$(DMET)_2Au(CN)_2$	20[3]	200	0.25	160	0.1[7]

	μ/μ_B Eq. (4.34)	$U(eV)$ Eq. (4.12)
$(TMTSF)_2PF_6$	0.01	2.0
$(DMET)_2Au(CN)_2$	0.008	2.0

1. D. Jerome and H. Schultz, Adv. Phys. **31**, 299 (1982).
2. T. Nakamura, et al. Solid State Comm. **75**, 583 (1990).
3. K. Kikuchi, et al. Jap. J. Appl. Phys. Supp. **26-3**, 1369 (1987).
4. T. Takahashi, et al. Physica **143B**, 417 (1986).
5. J. M. Delrieux, et al. Physica **143B**, 412 (1986).
6. K. Kanoda, et al. Phys. Rev. **B42**, 8678 (1990).
7. K. Kanoda, et al. Phys. Rev. **B38**, 39 (1988).

magnetic moments given in Table 4.1. The magnetic properties of the SDW state have not been calculated using the Hamiltonian Eq. (4.2). It is expected however, that the magnetic response and the collective excitations are close to those of an antiferromagnet; with localized spins which are a distance $\lambda_0/2$ apart and which correspond to a reduced moment μ and to an effective exchange constant J_{eff}. Therefore the Hamiltonian

$$\mathcal{H} = \left(\frac{\mu}{\mu_B}\right)^2 J_{\text{eff}} \sum_i \vec{S}_i \cdot \vec{S}_{i+1} - \left(\frac{\mu}{\mu_B}\right)^2 D^* \sum_i S_i^x S_{i+1}^x + \left(\frac{\mu}{\mu_B}\right)^2 E^* \sum_i (S_i^z S_{i+1}^z - S_i^y S_{i+1}^y) - g\mu \vec{H} \sum_i \vec{S}_i \tag{4.35}$$

has been used to evaluate the parameters which characterize the SDW ground state. Here J_{eff} is the interaction along the chains, and D^* and E^* represent the hard- and intermediate-axis anisotropy energies respectively; the reduced moment is written as μ/μ_B. The above Hamiltonian leads to an anisotropic susceptibility, with

$$\chi_{\parallel} \to 0 \tag{4.36a}$$

$$\chi_{\perp} = \frac{N_e g^2 \mu_0^2}{2\left(\frac{\mu_0}{\mu_B}\right)^2 J_{\text{eff}}} = \frac{N_e g^2 \mu_B^2}{2 J_{\text{eff}}} \tag{4.36b}$$

at zero temperature; the parallel direction refers to the magnetic field which is applied parallel to the easy axis along which the electron spins are aligned in the absence of an external magnetic field.

The effective exchange constant J_{eff} can be estimated using the following argument. In the spin density wave ground state a gap Δ develops for both spin subbands. However, there is no gap for spin-flip transitions, i.e., for transitions between the two spin subbands. In the presence of a magnetic field H, these subbands are shifted by $\pm\mu_B H$ and arguments applied for the Pauli susceptibility of the metallic state are appropriate. These lead to $\chi_{\perp}(T=0) = \chi_{\text{Pauli}}(T > T_{\text{SDW}})$ and in the absence of anisotropy effects the susceptibility is unaffected by the SDW transition. Then

$$\frac{g^2 \mu_B^2}{2 J_{\text{eff}}} N_e = \mu_B^2 n(\epsilon_F) \tag{4.37}$$

gives

$$J_{\text{eff}} = \frac{N_e}{n(\epsilon_F)} = \epsilon_F = \hbar v_F k_F. \tag{4.38}$$

With ϵ_F as given in Table 2.2 we expect J_{eff} to be on the order of 1000 K for the various materials.

The anisotropy as implied by Eq. (4.36) has been observed in $(TMTSF)_2X$ salts, and the susceptibility measured (Mortensen et al., 1981, 1982) in the AsF_6 salt is displayed in Fig. 4.3. The magnitude of $\chi_{\perp}$ leads, using Eq. (4.36a), to an effective coupling constant $J_{\text{eff}} = 740$ K. The Hamiltonian, Eq. (4.35), also leads to a spin-flop field given by

$$H_{sf} = \frac{\pi (2E^* J_{\text{eff}})^{1/2}}{\hbar \gamma} \tag{4.39}$$

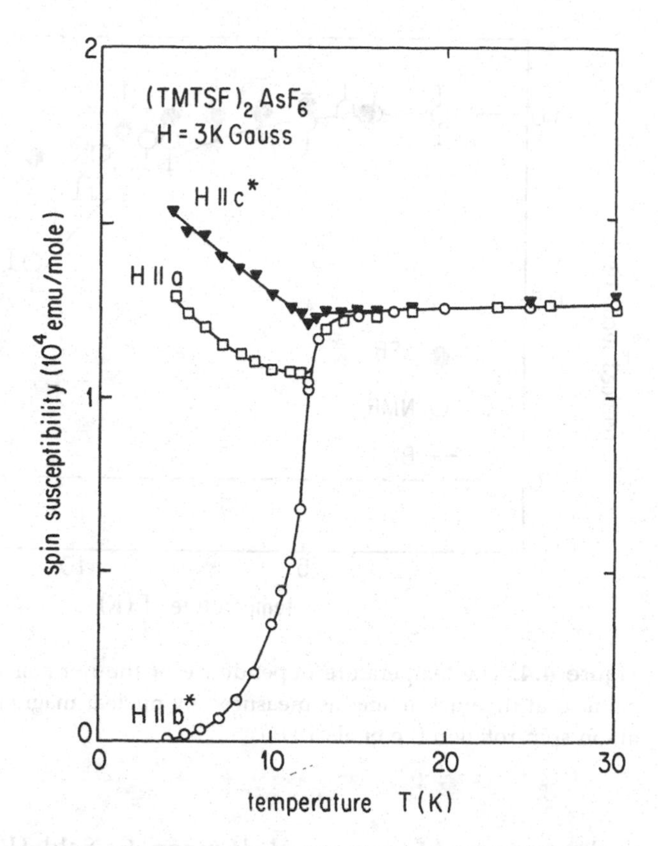

Figure 4.3. Temperature dependence of the single crystal spin susceptibility of $(TMTSF)_2AsF_6$ in a magnetic field (H = 3kOe) lower than the spin-flop field. The axis b^*, a, and c^* are respectively the easy, intermediate, and hard axis of the SDW state stable below 12 K (Mortensen et al., 1982).

where γ is the gyromagnetic ratio. Experiments at various magnetic fields, again in $(TMTSF)_2X$ salts (Mortensen et al., 1982), clearly establish the existence of a spin-flop field, which is on the order of $H_{sf} \simeq 6$ kG, from which an anisotropy energy of $E = 3 \times 10^{-5}$ K has been estimated.

The spatial variation of the spin density as given by Eq. (4.18) leads to a spatial variation of the internal magnetic field at the nuclear sites given by

$$\delta H(\vec{r}) = \langle a_0 \rangle H_0 \frac{\mu}{\mu_B} \cos\left(2\vec{Q} \cdot \vec{r} + \phi\right) \tag{4.40}$$

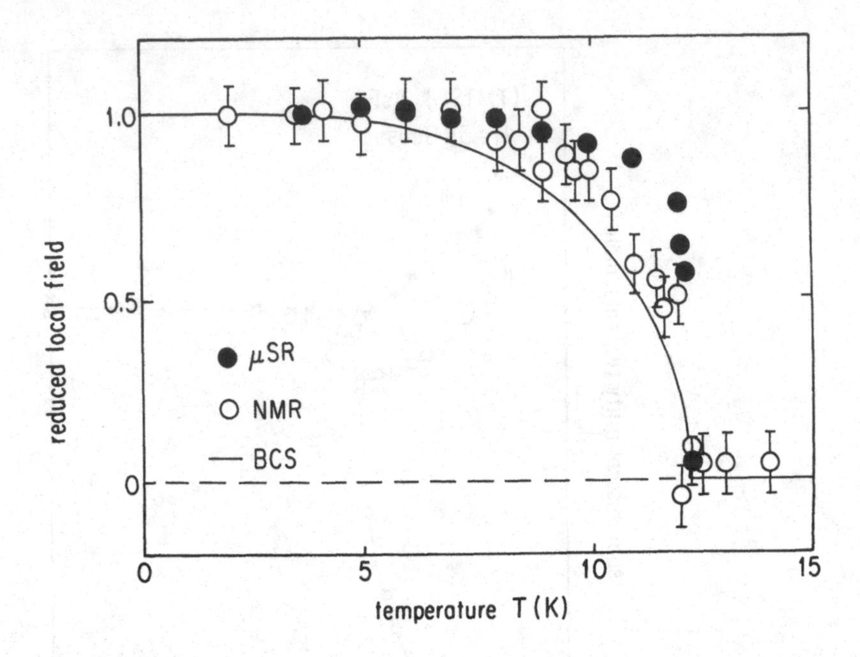

Figure 4.4. The temperature dependence of the normalized average internal field at the nuclear sites as measured by nuclear magnetic resonance and muon spin rotation (Le et al., 1991).

in the presence of an external *dc* magnetic field H_0 where $\langle a_0 \rangle$ is the hyperfine field interaction constant (a quantity which in general can be estimated). The internal field distribution, which is a sensitive function of the precise spatial dependence of the spin density, can be measured by local probes such as nuclear magnetic resonance (NMR) and muon spin rotation (μSR). Note that for a commensurate SDW, the distribution of the local fields at the nuclear sites is different from the distribution caused by an incommensurate SDW. Consequently, such studies can establish the incommensurate character of the density wave, and an elaborate analysis can also lead to the nesting wavevector, $\vec{Q} = (Q_\perp, 2k_F)$. This has been done for $(TMTSF)_2PF_6$, (Andrieux et al., 1981; Takahashi et al., 1986; Delrieux et al., 1981; Creuzet et al., 1982), leading to an SDW distortion wavevector (0.5, 0.24, 0.06) in units of a^*, b^*, and c^* (with b the chain direction); which is close to, but distinctively different from, a commensurate period $Q = (1/2, 1/4, 0)$. The period measured along the chain direction is also near the optimal nesting as arrived at by band structure calculations (Yamaji, 1982, 1983). The incommensurate SDW has also been established by μSR studies (Le et al., 1991). The magnitude of the magnetic field distribution, when measured at various temperatures, directly gives the temperature dependence of the order parameter and experimental results obtained both by NMR and μSR studies displayed in Fig. 4.4. The full line of the figure is the order parameter $\Delta(T)$, obtained from the weak coupling BCS theory. The somewhat stronger temperature dependence observed near T_{SDW} may be due to the weakly first order nature of the transition to the spin density wave ground state. The magnitude of the spin density wave modulation can also be extracted from the experimental results, leading to μ/μ_B which is close to that evaluated from NMR studies. This value is somewhat larger than the magnetic moment evaluated from the measured single particle gap and Coulomb interaction energy by using Eqs. (4.12), and (4.34).

Magnetic susceptibility and NMR studies have also been conducted in $(MDT\text{–}TTF)_2Au(CN)_2$ (Kanoda et al., 1990) and in $(DMET)_2Au(CN)_2$ (Kanoda et al., 1988). In both cases, the magnetic properties are similar to those of an antiferromagnet with a small magnetic moment; and the broadened NMR signal below T_{3D} clearly establishes that the spin density wave is incommensurate with the underlying lattice.

5 Fluctuation Effects

Whether we like it or not, whether we intend it or not, we are destined to have an authoritarian order.

—Alexander Solzenitsyn *Letters to the Soviet leaders*

As discussed in previous chapters, the mean field solution of one-dimensional models leads to finite transition temperatures T^{MF} where long range order develops. This is an artifact of the mean field approximation which neglects the role played by fluctuations of the order parameter. In strongly anisotropic materials, because of the reduction of the phase space such fluctuation effects become important; and a strictly one-dimensional system with short range interactions does not develop long range order at finite temperatures. Instead, only short range correlations occur at low temperatures (Landau and Lifschitz, 1958).

Of course, the materials which have been discussed are not strictly one-dimensional; Coulomb interactions between electrons on neighboring chains and also one-electron interchain tunneling lead to coupling of the fluctuations on neighboring chains. Such coupling leads again to a finite transition temperature, T_{3D}, below which three-dimensional long range order occurs. For weak interchain coupling, T_{3D} is significantly smaller than T^{MF}, the transition temperature given by the mean field approximation. The region below T^{MF} but above T_{3D} is characterized by one-dimensional fluctuations which, somewhat above T_{3D}, cross over to fluctuations which (depending on the anisotropy) have a two- or three-dimensional character. In addition to such effects the short coherence lengths observed, especially for charge density waves, also lead to fluctuations near T_{3D}, in much the same way that fluctuations arise in superconductors with short coherence lengths (Ginzburg, 1960).

Rigorous solutions for the behavior of the 1D chain with correlations which lead to a complex order parameter and also for the effects of interchain coupling are not available. The various approaches (Scalapino et al., 1975; Schultz, 1987), while leading to the same overall qualitative picture, give different expressions for the various characteristic temperatures and for the detailed temperature dependences of the experimentally accessible quantities. Here only a simplified description of the state of affairs is given, often referring to results which go well beyond the scope of these notes. The description which follows is appropriate for a set of weakly coupled chains where the mean field transition temperature is significantly higher than the temperature where three-dimensional long range order develops. This approach is appropriate for highly anisotropic materials; while for compounds for which the anisotropy of the electronic band structure is not large a different approach based on fluctuation theories in conventional three-dimensional solids (extended to anisotropic systems) may be more appropriate. Such an approach is summarized by Schultz (1987).

Experimental evidence for fluctuation effects has been found in a variety of materials in which charge or spin density wave ground states occur at low temperatures. Our understanding of fluctuations which precede the charge density wave transition are more complete, due partly to the observable structural changes which accompany such transitions; but also due to the availability of large single crystals on which the experiments can be performed. The most detailed structural, optical, and thermodynamic experiments have been conducted on $K_{0.3}MoO_3$, a material which undergoes a CDW transition at $T = 183$ K. Consequently, experiments on this compound will mainly be discussed and contrasted with the various models of fluctuations.

5.1 Fluctuations in Quasi-One-Dimensional Metals

5.1.1 Fluctuations for a One-Dimensional Chain

For a one-dimensional chain which is described by the Fröhlich Hamiltonian Eq. (3.11) or by the Hubbard Hamiltonian, Eq. (4.1), the order parameter is complex; and therefore when describing the temperature dependence of the correlation length, both the amplitude and phase fluctuations of the order parameter must be included. These fluctuations can be described by the Ginzburg-Landau functional, which for a one-dimensional chain is given as

$$F = F(0) + n(\epsilon_F) \int dx \left[a|\Delta|^2 + b|\Delta|^4 + c\left|\frac{d\Delta}{dx}\right|^2 \right] \tag{5.1}$$

where Δ is the complex order parameter.

The mean field description of the density wave transition was discussed in Chapters 3 and 4 in terms of the Landau theory; where the last term, representing the spatial fluctuations of the complex order parameter, is neglected. The shape of the free energy functional for a complex order parameter is shown in Fig. 5.1. At temperatures $T > T^{MF}$, with the mean field transition temperature as given by Eqs. (3.19) and (4.11b) for charge and spin density waves respectively; the amplitude of the order parameter fluctuates near $|\Delta| = 0$. The amplitude fluctuations are progressively removed as the temperature decreases, and at temperatures well below T^{MF}, the amplitude of the order parameter

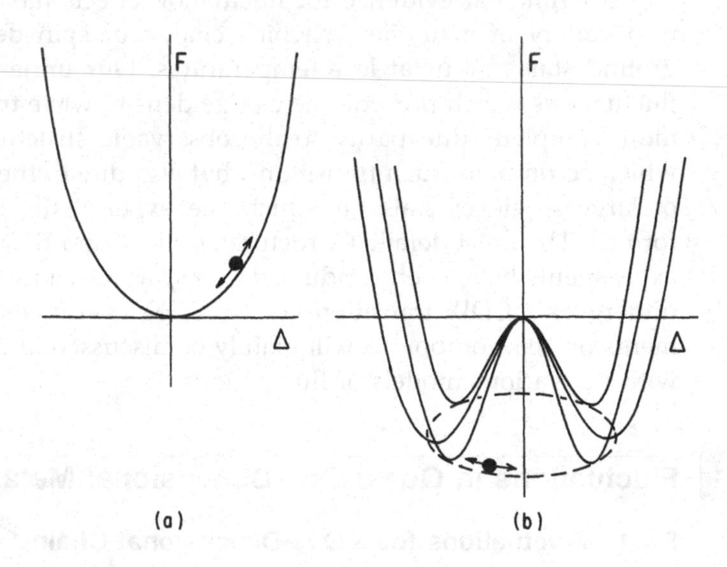

Figure 5.1. The free energy functional for a complex order parameter for $T > T_{MF}$ (a) and for $T < T_{MF}$ (b).

is close to its $T = 0$ value; and the phase of the order parameter rotates in the bottom of the potential well of $F(\Delta)$, as shown in Fig. 5.1. In this limit, the fluctuation of the order parameter is approximately given by

$$|\Delta|_0 \langle e^{i\phi(x)} \rangle \simeq |\Delta_0| e^{-\frac{1}{2}\langle \phi^2(x) \rangle} \tag{5.2}$$

where $|\Delta|_0$ is the magnitude of the gap at $T = 0$. For density waves the uniform phase $\phi = \text{const}$ corresponds to a frozen in lattice distortion with a wavevector $q = 2k_F$, and the spatial fluctuations of the phase are given in terms of the Fourier components as

$$\langle \phi^2(x) \rangle = \sum_{q-2k_F} \left\langle \phi^2_{q-2k_F} \right\rangle e^{i(q-2k_F)x} = \sum_{q'} \left\langle \phi^2_{q'} \right\rangle e^{iq'x}. \tag{5.3}$$

The energy associated with each component $\phi_{q'}$ is approximately given by $n(\epsilon_F)\delta^2(\epsilon) = n(\epsilon_F)[\hbar v_F q']^2$ for a one-dimensional band. From the equipartition theorem

$$n(\epsilon_F)(\hbar v_F q')^2 \langle \phi^2_{q'} \rangle = k_B T \tag{5.4}$$

which, together with (5.3), gives

$$\langle \phi^2(x) \rangle = \frac{\pi k_B T}{\hbar v_F} \int_0^{\pi/a} \frac{e^{iq'x}\, dq'}{(q')^2} \tag{5.5}$$

where we have replaced the summation with an integral over the Brillouin zone, and have used $n(\epsilon_F) = (\pi \hbar v_F)^{-1}$.

The integral on the right hand side diverges, and consequently, the order parameter as given by Eq. (5.2) has a vanishing expectation value at finite temperatures; hence there is no long range order.

The correlation function of the complex order parameter is given in this limit by

$$\langle \Delta^*(x), \Delta(0) \rangle = |\Delta|_0^2 e^{-\frac{1}{2}\langle [\phi(x) - \phi(0)]^2 \rangle}. \tag{5.6}$$

The spatial dependence of the phase is given by the terms of the Fourier components

$$\left\langle [\phi^2(x) - \phi^2(0)] \right\rangle = \sum_{q'} \phi^2_{q'} (e^{iq'x} - 1). \tag{5.7}$$

With the previous estimate of the energy associated with the fluctuating q component, we have

$$\langle[\phi^2(x)-\phi^2(0)]\rangle=\frac{\pi k_B T}{\hbar v_F}\int_0^{\pi/a}\frac{(1-\cos q'x)\,dq'}{(q')^2}=\frac{\pi k_B T}{\hbar v_F}x, \tag{5.8}$$

and consequently,

$$\langle\Delta^*(x),\Delta(0)\rangle=|\Delta|_0^2 e^{-\frac{\pi k_B T}{\hbar v_F}|x|}=e^{-\frac{|x|}{\xi_{1D}}} \tag{5.9}$$

with the phase-phase correlation length

$$\xi_{1D}=\frac{\hbar v_F}{\pi k_B T} \tag{5.10a}$$

which is the same as that obtained in Chapter 1 for the density fluctuations of the 1D electron gas. All this is appropriate well below T^{MF} where amplitude fluctuations can be neglected, and only phase fluctuations are important. At high temperatures, where the fluctuations of the amplitude are important, the correlation length is somewhat different and is given by (Lee, Rice, and Anderson, 1973)

$$\xi_{1D}=\frac{[7\zeta(3)]^{1/2}v_F}{4\pi k_B T} \tag{5.10b}$$

where $\zeta(3)$ is the third order zeta function. Consequently, the correlation length displays an approximate $1/T$ dependence (see Scalapino et al., 1975) at all temperatures with a change of the prefactor near T^{MF}.

The progressive development of density wave fluctuations is accompanied by corresponding changes in the phonon and electron densities of states. The susceptibility, which is related to the order parameter fluctuations, is defined as

$$\chi(q)=\frac{1}{k_B T}\int\langle\Delta^*(x),\Delta(0)\rangle e^{iqx}\,dx \tag{5.11}$$

and by virtue of the fluctuation dissipation theorem the static correlation function is given by

$$S(q)=k_B T\chi(q). \tag{5.12}$$

For charge density waves, the order parameter is related to

(or can be defined as) the periodic lattice distortion, and from Eq. (3.23)

$$S(q) = \frac{2\hbar g^2}{NM\omega_{2k_F}} \int \langle \Delta u(x), \Delta u(o) \rangle e^{iqx}\, dx. \qquad (5.13)$$

For spin density wave correlations, the order parameter is the staggered magnetization, and using Eq. (4.20)

$$S(q) = \frac{U^2}{N} \int \langle S(x), S(o) \rangle e^{iqx}\, dx. \qquad (5.14)$$

The fluctuations of the charge density wave order parameter can be examined by X-ray or neutron scattering. As there is no lattice distortion related to the development of spin density wave correlations, they can be studied only through magnetic neutron scattering. The scattering intensity is given by

$$I(k) = e^{-2W} \sum_H I\left(k - H_0 \cdot \vec{d} \pm 2k_F\right) S(q) \qquad (5.15)$$

where $\vec{d} = d_{\parallel}, d_{\perp}$ represents the lattice points, and $d_{\parallel}$ and $d_{\perp}$ are the lattice constants parallel and perpendicular to the chains, and H_0 is the Miller index of reflection. From Eq. (5.11)

$$\chi_{1D}(q_{\parallel}) = \frac{1}{k_B T} \int \langle |\Delta|^2 \rangle \exp\left[-\frac{x}{\xi_{1D}} \right] e^{iqx}\, dx \qquad (5.16)$$

$$= \frac{\langle |\Delta|^2 \rangle}{k_B T} \frac{\xi_{1D}/a}{1 + (q_{\parallel}\xi_{1D})^2}$$

where a is the lattice constant along the chain direction. This then, through Eqs. (5.12) and (5.13) leads to a broadened scattering peak, with a width determined by ξ_{1D}.

Long range order also leads to a well defined single particle gap, with a square root singularity at the gap which is characterized of a one-dimensional density of states as discussed in Chapter 3. Short range correlations lead instead to a pseudogap, with a suppressed density of states below the gap energy; the decrease of $D(\epsilon)$ for $\epsilon < |\Delta|$ becomes more pronounced for large correlation length. $D(\epsilon)$ has been calculated by Lee, Rice, and Anderson (1974) for a 1D Fröhlich Hamiltonian, and the results are shown

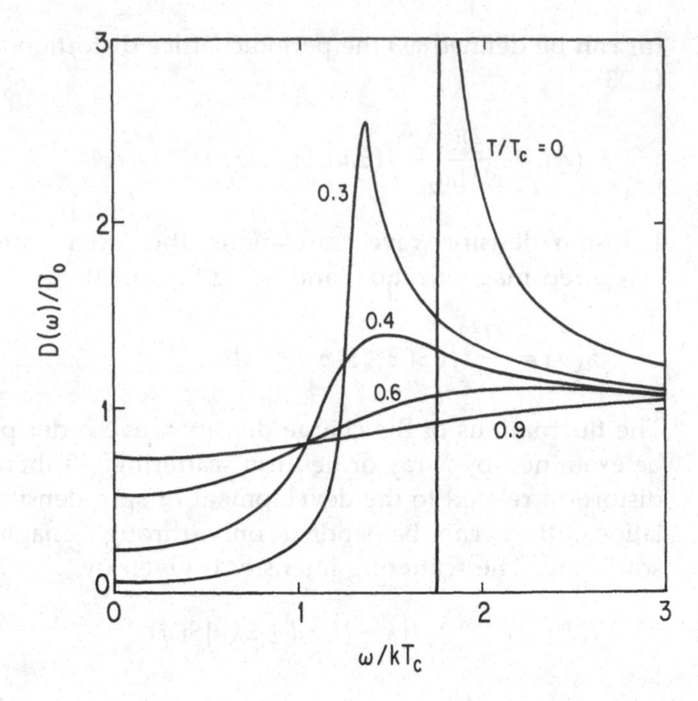

Figure 5.2. The density of states $D(\omega)$ for a one-dimensional chain undergoing a Peierls transition. T_c refers to the mean field transition temperature (Lee et al., 1973). D_0 is the (constant) density of states well above T_c.

in Fig. 5.2. In the Figure T_c refers to the mean field transition temperature, given in terms of the parameters of the model by Eq. (3.19).

5.1.2 Fluctuations for a System of Coupled Chains

The above picture is modified for a three-dimensional array of parallel chains. Due to interchain interactions density wave fluctuations on the neighboring chains become correlated and this leads to a transition to a ground state with three-dimensional, long range order. Let us assume a square lattice chains, with the lattice constant $d_\perp$. Then the interchain interaction can in general be written as

$$E_{\text{int}} = n(\epsilon_F)\frac{2}{d_\perp^2}c_\perp \sum_{i,j} dx\, Re\left[\Delta_i^*(x), \Delta_j(x)\right] \tag{5.17}$$

where i and j are the chain index and the summation is over nearest neighbors. In Eq. (5.17) we assume that the interchain

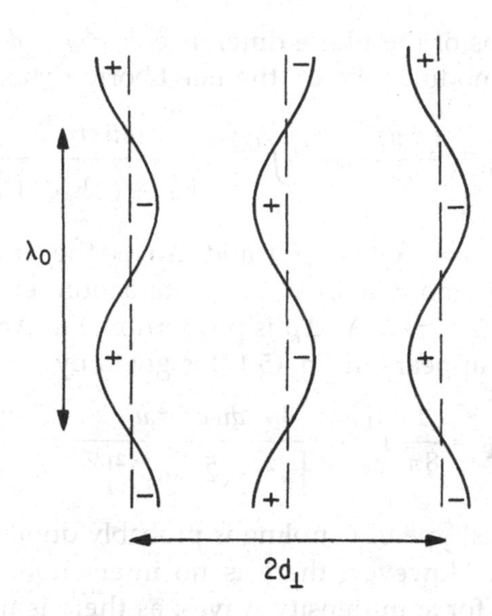

Figure 5.3. Charge density wave modulations on neighboring chains. Both Coulomb interactions and a quasi 2D Fermi surface lead to out of phase modulations of the density waves as shown in the figure.

interaction $c_\perp$ (and also the interchain distance) is the same in two directions perpendicular to the chain direction. In most cases $c_\perp > 0$, and E_{int} is minimum where the sign of the order parameter alternates between neighboring chains at distance $d_\perp$ apart (shown in Fig. 5.3). This alternation can be taken into account by defining a new order parameter

$$\Delta(\vec{r}) = (-1)^{(y_i + z_i)/d_\perp} \Delta_i(x). \tag{5.18}$$

With this definition the order parameter $\Delta(\vec{r})$ is constant in case of long range order and its fluctuations can be described within the framework of the Ginzburg-Landau theory, as will be done later.

Two mechanisms were found to be important in establishing an interaction between the neighboring chains. Coulomb interactions between the charge density wave modulations $\Delta\rho(x)$ on neighboring chains of unit length are given by (Saub et al., 1976)

$$V_{ij} = \frac{1}{4\pi} \int_0^1 dx_j \frac{\Delta\rho(x_i)\Delta\rho(x_j)}{\left[d_\perp^2 + (x_i - x_j)^2\right]^{1/2}}. \tag{5.19}$$

In terms of the phase difference $\phi = \phi_i - \phi_j$ of the density charge wave modulations on the neighboring chains,

$$V_{ij} = \frac{(\Delta\rho)^2}{8\pi}\cos\phi \int_0^{1/2k_F} \frac{du \cos^2 u}{\left[u^2 + (2k_F d_\perp)^2\right]^{1/2}} \tag{5.20}$$

where $u = 2k_F(x_i - x_j)$ and $\Delta\rho$ is the amplitude of the charge density modulations. The interaction energy has a minimum when $\phi = \pi/2$. As $\Delta\rho$ is proportional to $\Delta n(\epsilon_F)$, the coefficient $c_\perp$ which appears in Eq. (5.17) is given by

$$c_\perp = \frac{1}{8\pi}\int_0^{1/2k_F} \frac{du \cos^2 u}{\left[u^2 + (2k_F d_\perp)^2\right]^{1/2}}. \tag{5.21}$$

This type of coupling is probably dominant for charge density waves. However, there is no interaction arising from Coulomb effects for spin density waves; as there is no charge density wave modulation. Single particle tunneling between the chains also leads to ordering of density wave modulations on neighboring chains. This may be important both for charge and for spin density waves. The consequence of interchain tunneling is a two-dimensional (or quasi-three-dimensional) band structure, with a dispersion relation given by Eq. (1.12). The optimal nesting condition, discussed in Chapter 1, leads to alternating phase of the density modulations on the neighboring chains; which can be represented by an interchain coupling, the magnitude of which is given by (Horowitz et al., 1975).

$$c_\perp = 2\left(\frac{t_\perp d_\perp}{v_{F\parallel}}\right)^2 c. \tag{5.22}$$

where c is the Ginzburg-Landau coefficient defined in Eq. (5.1).

For $c_\perp$ as defined in Eq. (5.17) the wave vector dependent coupling constant is given by

$$c_\perp(q_\perp) = \frac{c_\perp}{2}\sum_j e^{iq_\perp d_\perp} = \frac{c_\perp}{2}\cos(q_\perp d_\perp). \tag{5.23}$$

This interchain coupling acts as an effective field, which tends to stabilize an ordered phase, with density wave modulation in opposite phase on neighboring chains. As usual the problem can be discussed within the framework of the mean field theory.

The susceptibility, $\chi(q_{\|}, q_{\perp})$ can be written in the mean field approximation as

$$\chi(q_{\|}, q_{\perp}) = \frac{\chi_{1D}(q_{\|})}{1 - Zc_{\perp}(0)\chi_{1D}(q_{\|})} \tag{5.24}$$

where Z is the number of nearest neighbors. The divergence of $\chi(q_{\|}, q_{\perp})$ defines a mean field transition temperature T_{3D} of the system of coupled chains, which is then given by

$$1 - Zc_{\perp}(0)\chi_{1D}(q_{\|})(T_{3D}) = 0 \tag{5.25}$$

As discussed earlier, the temperature dependence of ξ_{1D} and χ_{1D} have not been evaluated for a complex order parameter and the behavior is known only well below and well above T^{MF}. T_{3D} can therefore be evaluated only for certain assumptions on the temperature dependence of the coherence length. For weak interchain coupling the behavior well below T^{MF} is important. Then the amplitude of the order parameter is close to its $T = 0$ value, and interchain coupling leads to an adjustment of the phase of the density wave modulations on the neighboring chains. The transition temperature can then be estimated using the following simple argument. Correlations over the coherence length $\xi_{1D}(T)$ are coupled due to interchain interactions, and the energy gain due to such phase adjustment is on the order of

$$\Delta E_{\perp} = Zc_{\perp}\, n(\epsilon_F)|\Delta(T = 0)|^2 \xi_{1D}(T). \tag{5.26}$$

This leads to a transition temperature T_{3D} which, in the spirit of the mean field approximation, is approximately given by

$$k_B T_{3D} \simeq \Delta E_{\perp}. \tag{5.27}$$

By virtue of Eq. (5.10a) we obtain

$$k_B T_{3D} = \left[\pi^{-1} Zc_{\perp} n(\epsilon_F) |\Delta(T = 0)|^2 \hbar v_F\right]^{1/2} \tag{5.28}$$

in the limit where $T_{3D} \ll T^{MF}$ and only phase fluctuations play an important role. The same transition temperature is obtained using Eq. (5.25), with ξ_{1D} as given by Eq. (5.16).

In the case where the fluctuations of the amplitude are important and Eq. (5.10b) applies, the same relation, with a prefactor different from π^{-1}, is obtained. The transition temperature, T_{3D}, where three-dimensional order develops is, for a small interchain coupling $c_\perp$, significantly smaller than T_{MF} as given by Eq. (3.19); and the transition temperature approaches zero for vanishing interchain coupling, as expected.

Having evaluated the mean field transition temperature where 3D order develops; the behavior near T_{3D} can be discussed using the Ginzburg-Landau functional, and Eq. (5.4) together with the interchain coupling term, Eq. (5.17), becomes

$$F = F(0) + \frac{1}{d_\perp^2} n(\epsilon_F) \int d^3\vec{r} \tag{5.29}$$

$$\times \left\{ a|\Delta|^2 + b|\Delta|^4 + c\left|\frac{d\Delta}{dx}\right|^2 + c_\perp^* |\nabla_\perp \Delta(\vec{r})|^2 \right\}$$

where $c_\perp^* = c_\perp d_\perp^2$, and $\nabla_\perp$ is the gradient with respect to the transverse spatial variables. The spatially varying order parameter is written as

$$\Delta(x) = \sum_k \Delta_k e^{ikx} \tag{5.30}$$

and, in terms of Δ_k, F is given by

$$F = F(0) + \sum a_k |\Delta_k|^2 + \frac{b}{2} \sum_{k_1 k_2 k_3} \Delta_{k_1}^* \Delta_{k_2}^* \Delta_{k_3} \Delta_{k_1+k_2-k_3} \tag{5.31}$$

with

$$a_k = a + ck_x^2 + c_\perp \left(k_y^2 + k_z^2\right). \tag{5.32}$$

If we neglect the last term in Eq. (5.31), the correlation function of the order parameter is given by

$$\langle \Delta^*(\vec{r}), \Delta(o) \rangle = \frac{1}{2\pi} \int d\vec{k} \langle |\Delta_k|^2 \rangle e^{i\vec{k}\cdot\vec{r}}. \tag{5.33}$$

From the fluctuation-dissipation theorem (note that here we assume that in 3D there is a phase transition, and a_k is finite)

$$\langle |\Delta_k|^2 \rangle = \frac{k_B T}{a_k} \tag{5.34}$$

and with Eq. (5.32) the integral in Eq. (5.33) can be evaluated in three dimensions leading to

$$\langle \Delta^*(\vec{r}), \Delta(o) \rangle \simeq \frac{1}{\left[\left(\frac{x}{\xi_\parallel}\right)^2 + \left(\frac{r_\perp}{\xi_\perp}\right)^2\right]^{1/2}} e^{-\left[\left(\frac{x}{\xi_\parallel}\right)^2 + \left(\frac{r_\perp}{\xi_\perp}\right)^2\right]^{1/2}} \tag{5.35}$$

with the two correlation lengths

$$\xi_\parallel = \left(\frac{c}{a}\right)^{1/2} \tag{5.36a}$$

$$\xi_\perp = \left(\frac{c_\perp}{a}\right)^{1/2}. \tag{5.36b}$$

In the spirit of the Gaussian approximation we assume that near T_{3D}, $a = a'(T - T_{3D})$, and therefore

$$\xi_\parallel = \left(\frac{c}{a'}\right)^{1/2} (T - T_{3D})^{-1/2}. \tag{5.37a}$$

$$\xi_\perp = \left(\frac{c_\perp}{a'}\right)^{1/2} (T - T_{3D})^{-1/2}. \tag{5.37b}$$

Thus, both the on-chain and perpendicular correlation lengths diverge at T_{3D}, and the square root behavior of the coherence length is typical of the Ginzburg-Landau treatment in the Gaussian approximation of the fluctuations, where the last term in Eq. (5.31) is neglected. Near T_{3D} both $\xi_\parallel$ and $\xi_\perp$ are large, and $\xi_\perp$ exceeds $d_\perp$, the distance between the chains. In this region, the fluctuations have a three-dimensional character, with fluctuations on neighboring chains strongly coupled. With increasing temperature $\xi_\perp$ decreases, and at high temperatures it becomes smaller than $d_\perp$. In this temperature regime fluctuations on neighboring chains are decoupled and have a one-dimensional character. From (5.37b) this crossover from 3D to 1D fluctuations happens when

$$d_\perp = \left(\frac{c_\perp}{a'}\right)(T^* - T_{3D})^{-1/2}. \tag{5.38}$$

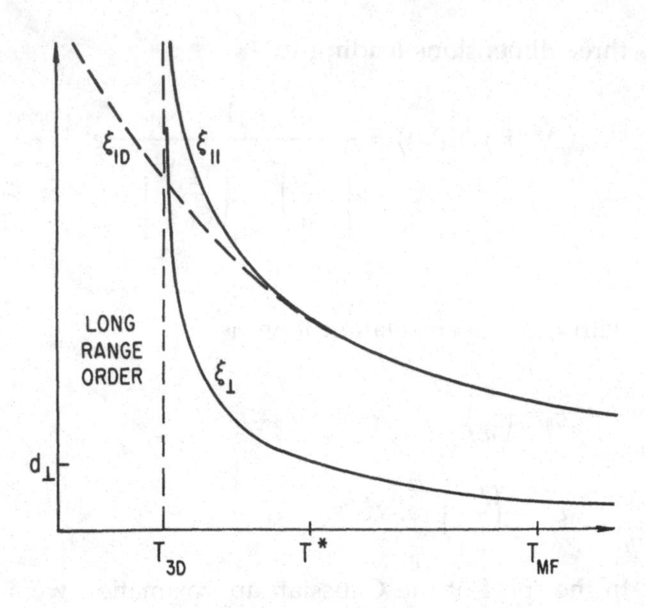

Figure 5.4. The temperature dependence of the correlation lengths parallel and perpendicular to the chains. The temperature dependence of the correlation length of a 1D chain (given by Eqs. (5.10a) and (5.10b) at low and at high temperatures) is also shown. The interchain lattice constant, $d_{\perp}$, is also indicated. T_{MF} is the mean field transition temperature, and there is a 1D to 3D crossover for fluctuations at T^*.

This then defines a crossover temperature

$$T^* = T_{3D} + \frac{d_{\perp}^2 a'}{c_{\perp}} \tag{5.39}$$

The temperature dependences of $\xi_{\parallel}$ and $\xi_{\perp}$ are displayed in Fig. 5.4, together with ξ_{1D} given by Eqs. (5.9) and (5.10) well below and well above T_{MF}.

The picture which emerges from the above analysis is as follows. At temperatures near T_{MF}, the system of coupled chains does not undergo a phase transition; instead correlations build up along the chains with the correlation length given by Eq. (5.10). Because the $2k_F$ fluctuations on the neighboring chains are not correlated the scattering intensity $I(k)$ is essentially that of a 1D chain: the scattering cross section is that of parallel lines at a distance $2k_F$ from the Bragg diffraction spots. The situation, appropriate for a two-dimensional array of chains is shown in Fig. 5.5. Below approximately T^*, correlations among the fluctua-

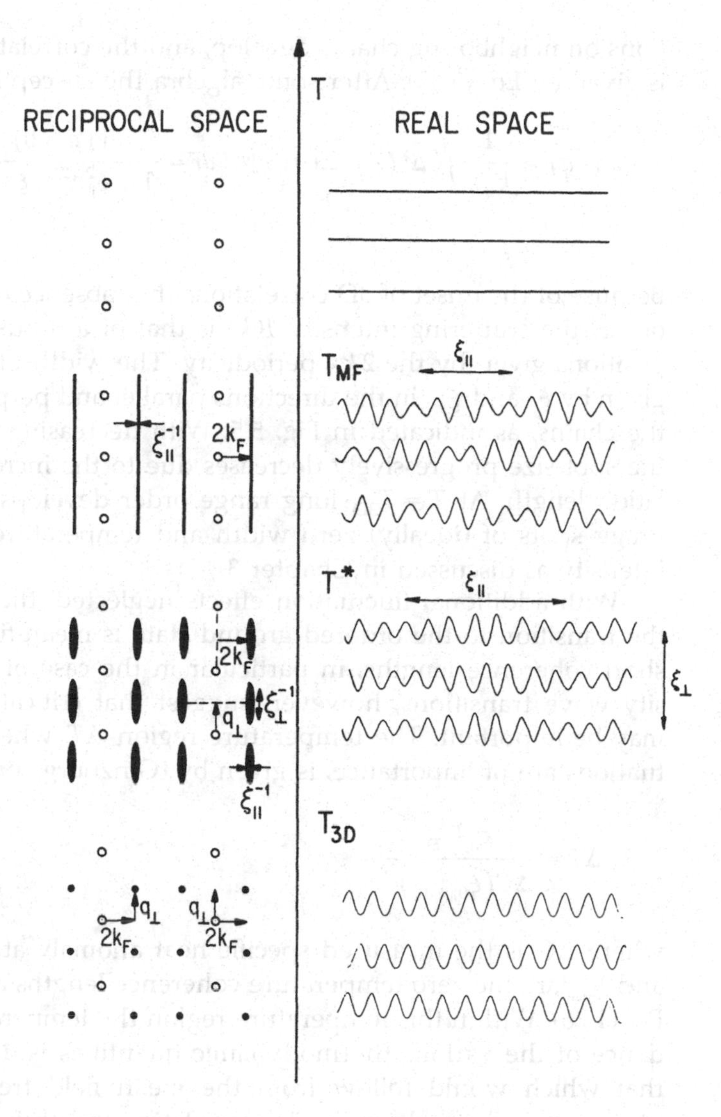

Figure 5.5. Fluctuations in a system of coupled chains in the 1D and 3D fluctuation regime and the corresponding long range correlations below T_{3D}. The main features of the diffraction pattern are also displayed in the different temperature regions.

tions on neighboring chains develop, and the correlation function is given by Eq. (5.35). After some algebra the susceptibility

$$\chi(q) = \frac{1}{k_B T}\int \langle \Delta^*(\vec{r}), \Delta(0)\rangle e^{i\vec{q}\cdot\vec{r}}\, d\vec{r} = \frac{\chi(q=0)}{1+\xi_{\parallel}^2 q_{\parallel}^2 + \xi_{\perp}^2 q_{\perp}^2}. \tag{5.40}$$

Because of the onset of 3D correlations (but absence of long range order) the scattering intensity $I(k)$ is that of a diffuse spot, with positions given by the $2k_F$ periodicity. The width of the spots is given by $\xi_{\parallel}$ and $\xi_{\perp}$ in the directions parallel and perpendicular to the chains, as indicated in Fig. 5.5. With decreasing temperature the spot size progressively decreases due to the increasing correlation length. At $T = T_{3D}$ long range order develops resulting in Bragg spots of (ideally) zero width and temperature dependent intensity as discussed in Chapter 3.

With additional fluctuation effects neglected, the behavior of the transition to the ordered ground state is mean-field-like. The short coherence lengths, in particular in the case of charge density wave transitions, however, suggest that critical fluctuations may be important. The temperature region ΔT where such fluctuations are of importance, is given by (Ginzburg, 1960)

$$\Delta T = \frac{k_B^2 T_{3D}}{\Delta C (\xi_{0\parallel}\xi_{0\perp})^2} \tag{5.41}$$

where ΔC is the measured specific heat anomaly at T_{3D} and $\xi_{0\parallel}$ and $\xi_{0\perp}$ are the zero temperature coherence lengths as defined by Eq. (1.26). Within this temperature region the temperature dependence of the various thermodynamic quantities is different from that which would follow from the mean field treatment. The temperature dependence of the correlation length is given by

$$\xi^{-1} \sim (T - T_{3D})^{\nu} \tag{5.42}$$

where the exponent depends on the dimensionality of the fluctuations. For three dimensions, and for a complex order parameter, the exponent $\nu = 0.67$ (Dietrich, 1976). The exponent ν is different from $\nu = 1/2$, which is appropriate for a mean field transition. This leads to a structure factor and to thermodynamic quantities which have a temperature dependence distinctly different from those which are given by the mean field approximation.

5.2 Charge Density Wave Fluctuations in $K_{0.3}MoO_3$

The linear chain compound $K_{0.3}MoO_3$ undergoes a metal-insulator transition at $T_{3D} = 183$ K, and various experiments discussed in Chapter 3 give clear evidence for a charge density wave ground state below the temperature where long range order develops. Because of the availability of large single crystals, a wide range of transport, thermodynamic, and scattering experiments have been performed at temperatures above T_{3D}, and the important role of fluctuations has been established.

The single particle gap is $\Delta \simeq 700$ K at $T = 0$ (see Table 3.1), and this value, together with the weak coupling expression of the mean field transition temperature $2\Delta = 3.5 k_B T^{MF}_{CDW}$, leads to $T^{MF}_{CDW} = 330$ K; well above the temperature where long range order due to the coupling between the neighboring chains develops. There is therefore a large critical region where fluctuation effects are expected to be important.

These fluctuations have been examined in detail by neutron scattering measurements of the structure factor $S(q)$. Scans in different directions, when analyzed in terms of Eqs. (5.12) and (5.40), lead at room temperature (i.e., somewhat below T^{MF}_{CDW} but well above T_{3D}), to a behavior close to that shown in the upper half of Fig. 5.5. The correlation length along the chains is $\xi_{\parallel} = 20$ Å, exceeding the lattice constant, while in the directions perpendicular to the chains $\xi_{\perp} < d_{\perp}$; and consequently the fluctuations on the neighboring chains are decoupled. The correlation lengths increase with decreasing temperature and at $T^* \simeq 200$ K, $\xi_{\perp}$ measured for both directions perpendicular to the chains exceeds $d_{\perp}$. Therefore at approximately this temperature a smooth transition to the region with 3D fluctuations occurs. The temperature dependence of the inverse coherence length, measured in the various crystallographic directions (parallel to the chains and in the two perpendicular directions), is shown in Fig. 5.6. The full line, which gives an appropriate representation of the experimental results of Eq. (5.42) with $\nu = 0.68$; is in excellent agreement with that expected for the $D = 3$, $n = 2$ universality class for a

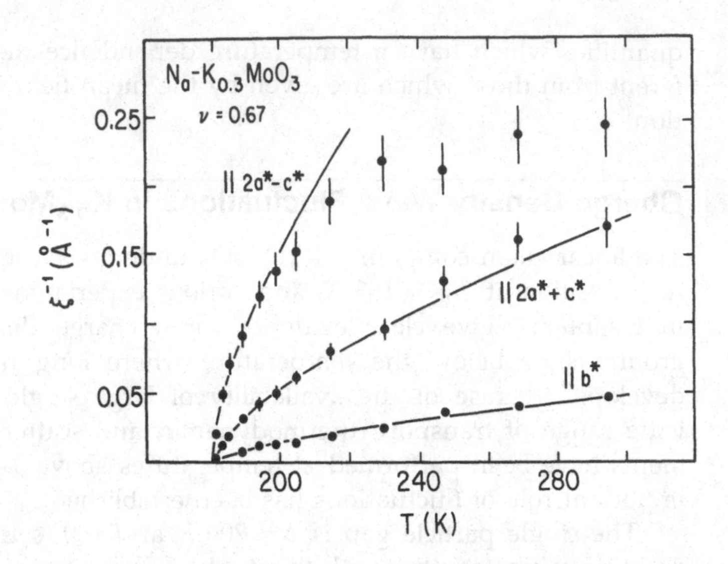

Figure 5.6. Temperature dependences of the inverse correlation lengths, measured in $K_{0.3}MoO_3$ in three different directions. The full line is Eq. (5.42), near T_{3D} with $\nu = 2/3$. (Pouget, 1989; Pouget and Comes, 1989 and references cited therein). The mean field result is $\nu = 1/2$.

transition to an ordered state characterized by a complex order parameter. These experiments suggest that the Ginzburg critical region is large and ΔT is on the order of 20 K. This is not unreasonable, since the measured specific heat anomaly of $\Delta C = 3.6$ J(mole K)$^{-1}$ and the coherence volume $\xi_{\|}\xi_{\perp 1}\xi_{\perp 2} =$ 420 Å^3 (Girault et al., 1989), indeed give $\Delta T = 20$ K. While the above discussion leads to well defined temperature ranges when one- or three-dimensional fluctuations occur, the actual state of affairs is certainly more complicated, as a description in terms of an isotropic interchain coupling $c_{\perp}$ is an oversimplification and the interchain coupling is different in the two crystallographic directions perpendicular to the chain direction. Consequently, a more detailed comparison between theory and experiment is not justified.

Fluctuation effects also lead to pronounced changes in the density of single particle electron states, and the gradual opening of the pseudogap has been inferred from the magnetic susceptibility and optical properties. The susceptibility $\chi(T)$ displayed in Fig. 5.7, first slowly decreases with decreasing temperature in

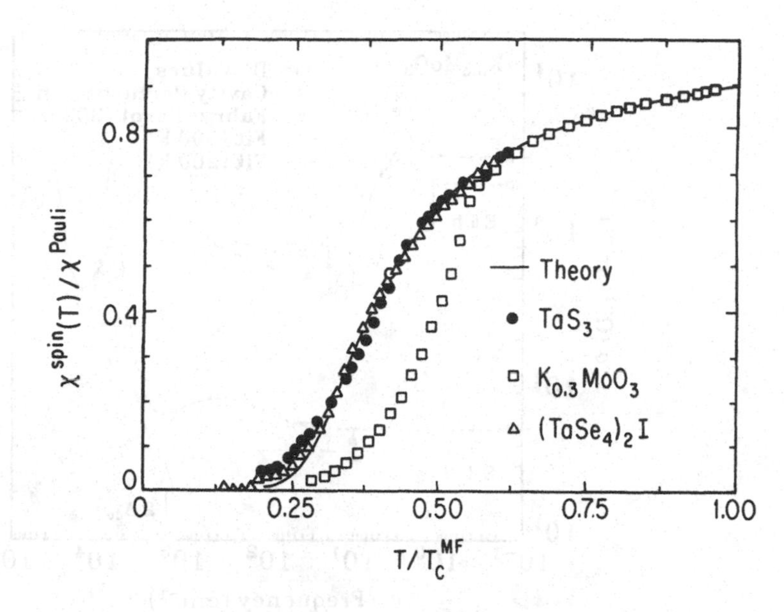

Figure 5.7. The temperature dependence of the magnetic susceptibility in $K_{0.3}MoO_3$ and other compounds with CDW ground states. (Johnston et al., 1985). The full line is the result of a model calculation based on the theory of Lee et al. (1973).

the region near T_{CDW}^{MF} where the fluctuations of the amplitude are important. The 3D transition temperatures are displayed in Table 3.1. Then using 2Δ (also given in Table 3.1) with Eq. (3.46) to evaluate T_{MF} gives the T_{3D}/T_{MF} ratios 0,4; 0,6; and 0,26 for TaS_3, $K_{0.3}MoO_3$, and $(TaSe_4)_2I$. The rapid decrease somewhat above T_{3D} is due to the more pronounced development of the pseudogap, with phase fluctuations of primary importance. The full line on the figure is the result of a model calculation, based on the theory of Lee, Rice, and Anderson (1973), which also leads to the density of states as shown in Fig. 5.2.

The gradual opening of the pseudogap has been directly confirmed by optical studies (Degiorgi and Grüner, 1992, Gorshunov et al., to be published), and the optical conductivity, measured at two different temperatures above T_{3D}, is shown in Fig. 5.8. The conductivity is depressed in the range between 100 and 10^3 cm^{-1}; indicating a pseudogap with decreasing temperatures, in full qualitative agreement with theory, worked out by

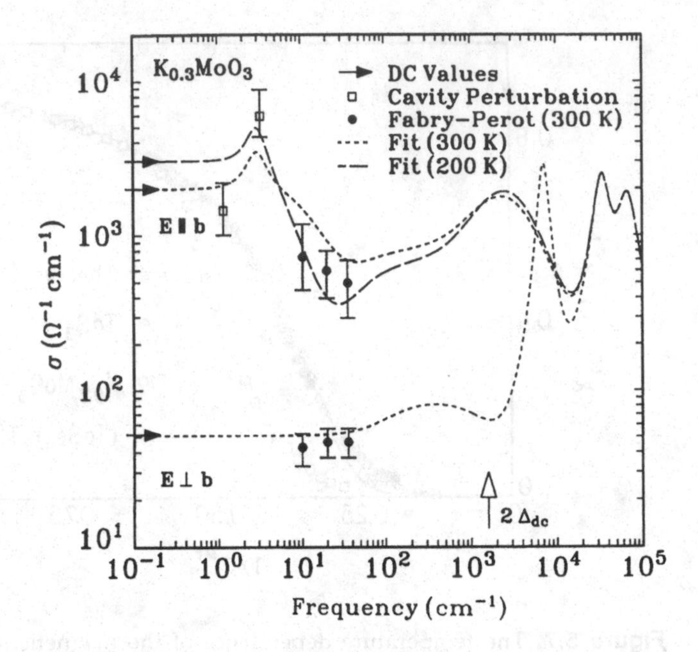

Figure 5.8. Optical conductivity measured at two different temperatures above $T_{3D} = 180$ K in $K_{0.3}MoO_3$ both parallel and perpendicular to the chain direction b. The notation "Fit" refers to the Kramers-Kronig type analysis of the reflectivity (Gorshunov et al., to be published).

Lee, Rice, and Anderson (1973). There is, however, a narrow peak near $\omega = 0$ along the chain direction, which becomes progressively more pronounced as T_{3D} is approached from above. This most likely is due to the fluctuating CDW segments which contribute to the conductivity. A detailed theory however, is not available to date.

The various thermodynamic quantities, such as specific heat, thermal expansion, and elastic properties are dominated, in the temperature range well above T_{3D}, by the (temperature dependent) contributions due to the underlying lattice. Consequently, such experiments do not give information on low-dimensional fluctuation effects. These quantities, however, have been explored at temperatures near T_{3D} and they give clear evidence for critical fluctuations and for deviations from the mean field approximation. The specific heat (Kwok and Brown, 1989; Kwok et al., 1990), thermal expansion (Hauser et al., 1991), and elastic properties

(Aronowitz et al., 1990) all show significant deviations from the mean field behavior and are suggestive of a critical region of significant magnitude. The detailed behavior near T_{3D} depends on the available degrees of freedom and on dimensionality. As the charge density wave has two degrees of freedom, amplitude and phase, it belongs to the same universality class as the *XY* model. This analogy has also been widely used to analyze the various experimental results near the transition temperature, and some of the critical exponents have also been extracted (Aronowitz et al., 1990).

6 Collective Excitations

...not from any one thing in particular but from myriad of voices comprising a chorus of promise.

—Czeslaw Milosz *The Issa valley*

The excitations of the density wave states which are related to the spatially and time-dependent order parameter $\Delta(\vec{r}, t)$ are difficult to describe within the framework of a microscopic description, and therefore are usually treated by using the time-dependent Ginzburg-Landau theory in the long wavelength limit. As expected for a complex order parameter, both phase and amplitude excitations occur. To the first order the modes are decoupled and represent independent oscillations of the amplitude and phase of the order parameter.

In the $q = 0$ limit the phase excitation corresponds to the translational motion of the undistorted condensate. For an incommensurate density wave such translational motion does not change the condensation energy, and consequently the frequency of the phase mode is

$$\omega_\phi(q = 0) = 0. \tag{6.1}$$

Such translational motion, nevertheless, involves ionic motions in the case of charge density waves. This leads to an enhanced kinetic energy and hence to a large effective mass. Mass enhancement does not occur for spin density waves which do not couple to the underlying lattice.

In the $q = 0$ limit, the frequency of the oscillations of the amplitude mode is the same as the gap frequency for spin density waves. For a charge density wave ground state fluctuations of the single particle gap $\delta\Delta$ with a frequency ω_A also lead to fluctuations of the ionic positions $\delta(\Delta u)$. This in turn leads to an oscillation frequency different from $\Delta/\hbar$. In the $q = 0$ limit, a $\delta\Delta$ change of the gap gives a change of the condensation energy $\frac{1}{2}n(\epsilon_F)(\delta\Delta)^2$; while the kinetic energy associated with the ionic displacements is given by $\frac{1}{2}MN_e\omega_A^2(q=0)\{\delta(\Delta u)^2\}$ where Δu is the atomic displacement. Therefore

$$\frac{1}{2}MN_e\omega_A^2(q=0)\left\{\delta(\Delta u)^2\right\} = \frac{1}{2}n(\epsilon_F)(\delta\Delta)^2 \tag{6.2}$$

and

$$\omega_A^2(q=0) = \frac{n(\epsilon_F)(\delta\Delta)^2}{MN_e\{\delta(\Delta u)\}^2}. \tag{6.3}$$

Assuming that $\delta(\Delta u)/\Delta u = \delta\Delta/\Delta$, and using Eqs. (3.20) and (3.23), we obtain

$$\omega_A^2(q=0) = \lambda^{1/2}\omega_{2k_F}. \tag{6.4}$$

Generally $\omega_{2k_F} < \Delta$ and $\lambda < 1$ for most materials with a charge density wave ground state; the amplitude mode frequency is well below the continuum of single particle excitations. Both modes are shown in the $q = 0$ limit in Fig. 6.1. The oscillations of the phase involve the displacement of electronic charge distributions with respect to the ionic positions; and consequently this mode is optically active. Such displacements do not occur for amplitude fluctuations and therefore the amplitude mode is expected to be Raman active.

In this chapter we first discuss these collective excitations for a one-dimensional chain with a charge density wave ground state and derive the dispersion relation of the excitations in terms of the parameters of the Ginzburg-Landau theory. These parameters can be related to the underlying microscopic theory in certain limiting cases. Consequently the dispersion relations can then be described in terms of the parameters of the microscopic theory.

The excitations of the spin density wave ground state are more complex, as both charge and spin degrees of freedom are important. The former are similar to excitations of the charge density wave state. The spin excitations are to some extent similar to excitations of an antiferromagnet; with the dispersion relation

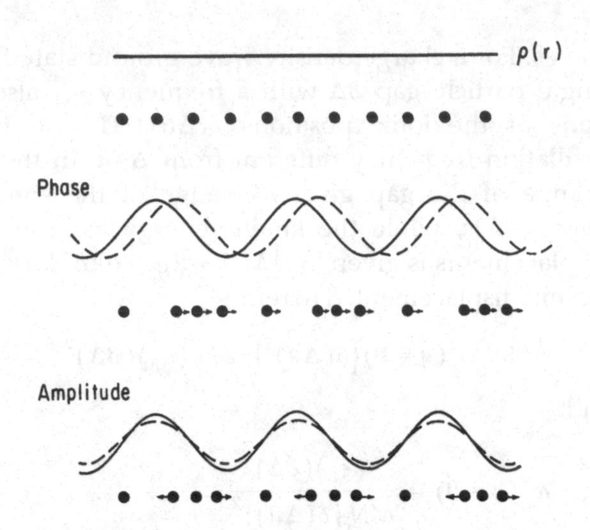

Figure 6.1. Amplitude (A) and phase (ϕ) excitations of the charge density wave state in the $q = 0$ limit. Changes of both the charge density and ionic displacements are indicated. The upper part of the figure shows the charge density and the atomic positions in the metallic state.

determined by the parameters, such as the effective exchange constant J_{eff}, and effective magnetic moment μ.

6.1 Ginzburg-Landau Theory of Charge Density Wave Excitations

We write the free energy of a one-dimensional chain as

$$F = F(0) + n(\epsilon_F) \int dx \left\{ a|\Delta|^2 + b|\Delta|^4 + c\left|\frac{d\Delta}{dx}\right|^2 + d\left|\frac{d\Delta}{dt}\right|^2 \right\}. \tag{6.5}$$

The third term on the right hand side describes the energy associated with the spatial variation; the last term represents the kinetic energy, due the temporal fluctuations of the order parameter.

We assume that the time and spatially dependent order parameter can be written as

$$\Delta(x,t) = [|\Delta| + \delta(x,t)]e^{i\phi(x,t)}. \tag{6.6}$$

Here $|\Delta|$ is the magnitude of the spatial and temporal averaged gap, $\delta(x,t)$ describes the fluctuation of the amplitude of the order parameter, and $\phi(x,t)$ refers to phase fluctuations. Within the framework of this approximation, two modes, an amplitude mode and a phase mode, represent the long wavelength excitations of the condensate. Expanding the expression of the free energy for small δ and ϕ we obtain

$$F = F(0) + n(\epsilon_F)\int dx\left\{a|\Delta|^2 + a\delta^2 + c\left(\frac{\partial\delta}{\partial x}\right)^2\right. \tag{6.7a}$$

$$\left. + d\left(\frac{\partial\delta}{\partial t}\right)^2 + \Delta^2\left[c\left(\frac{\partial\phi}{\partial x}\right)^2 + d\left(\frac{\partial\phi}{\partial t}\right)^2\right]\right\}.$$

The Lagrangian density which corresponds to the free energy is given by

$$\mathscr{L} = d\left(\frac{d\Delta}{dt}\right)^2 - c\left(\frac{d\Delta}{dx}\right)^2 - a\Delta^2 \tag{6.7b}$$

and the equations of motion for the amplitude and phase fluctuations are determined by the Lagrange equation

$$\frac{\partial\mathscr{L}}{\partial\phi} - \frac{\partial}{\partial t}\frac{\partial\mathscr{L}}{\partial\phi_t} - \frac{\partial\mathscr{L}}{\partial x}\frac{\partial\mathscr{L}}{\partial\phi_x} = 0 \tag{6.8a}$$

and

$$\frac{\partial\mathscr{L}}{\partial\delta} - \frac{\partial}{\partial t}\frac{\partial\mathscr{L}}{\partial\delta_t} - \frac{\partial\mathscr{L}}{\partial x}\frac{\partial\delta}{\partial\delta_x} = 0, \tag{6.8b}$$

where the subscripts refer to the time and spatial derivatives of ϕ and δ. Then with Eq. (6.7) we obtain

$$d\frac{d^2\phi}{dt^2} - c\frac{d^2\phi}{dx^2} = 0 \tag{6.9a}$$

$$d\frac{d^2\delta}{dt^2} - c\frac{d^2\delta}{dx^2} + 2a\delta^2 = 0. \tag{6.9b}$$

We seek the wave-like solutions for the amplitude and phase mode

$$u_A = u_{A_0}e^{i(\omega t - qx)} \tag{6.10a}$$

$$u_\phi = u_{\phi_0}e^{i(\omega t - qx)}. \tag{6.10b}$$

Where the displacement u can be parallel or perpendicular to the direction of the wave propagation. The former represents a longitudinal mode, the latter a transverse mode. The dispersion relations for the amplitude and phase oscillations are given by

$$\omega_\phi(q) = \left(\frac{c}{d}\right)^{1/2} q \tag{6.11a}$$

and

$$\omega_A = \left(\frac{-2a}{d} + \frac{c}{d} q^2\right)^{1/2}. \tag{6.11b}$$

The excitations of the phase and amplitude modes are usually referred to as phasons and amplitudons, respectively. The parameters of the Ginzburg-Landau free energy can be derived in terms of the microscopic description of the charge density wave ground state, by calculating the spatial and temporal fluctuations using an appropriate decoupling procedure. Instead of such general treatment, only the long wavelength fluctuations which are associated with variations of the phase will be discussed; in the limit where the order parameter is close to its $T = 0$ value. Then the spatial and temporal fluctuations of the order parameter are given to the first order as

$$\left(\frac{d\Delta}{dx}\right)^2 \simeq |\Delta|^2 \left(\frac{d\phi}{dx}\right)^2 \tag{6.12a}$$

$$\left(\frac{d\Delta}{dt}\right)^2 \simeq |\Delta|^2 \left(\frac{d\phi}{dt}\right)^2. \tag{6.12b}$$

Such an approach is appropriate well below T_{MF} where the amplitude and phase of the order parameter can clearly be distinguished. Assume a smooth local variation of the phase, with $d\phi/dx$ constant over many CDW wavelengths. This, by virtue of $2\,\delta k x = \delta\phi$, leads to a change of the wavevector by $\delta k = \frac{1}{2} d\phi/dx \,\mathrm{sign}(k)$ at the two sides of the Fermi surface. Therefore, there is a shift of the single particle gap from $\pm k_F$ to $\pm(k_F + \frac{1}{2} d\phi/dx)$ as shown in Fig. 6.2a. If the gap follows the spatial fluctuations and is constrained to be at the Fermi level, the electronic density must be changed, and this is described by

$$\rho(x) = \rho_0 + \frac{1}{\pi}\frac{d\phi}{dx}. \tag{6.13}$$

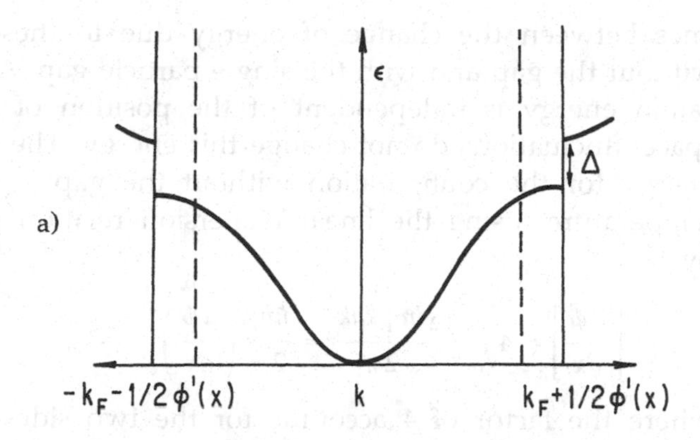

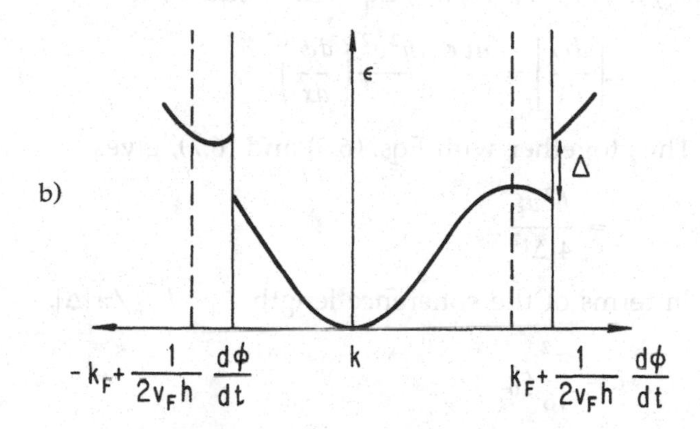

Figure 6.2. The displaced Fermi surface and the dispersion election in the case of a position- and a time-dependent phase. The states are filled up to the Fermi level, $\pm[k_F + \frac{1}{2}(d\phi(x)/dx)]$ in (a) and up to $\pm[k_F \pm (1/2v_F\hbar)(d\phi/dt)]$ in (b). The dotted lines refer to the Fermi surface with a position- and time-independent phase, and are at the positions $\pm k_F$ in both figures.

The energy associated with the spatially uniform phase is the energy difference between the state with and without the formation of the gap. At $T = 0$ this is simply the condensation energy $E = \frac{1}{2}n(\epsilon_F)\Delta^2$, and consequently,

$$a = -\tfrac{1}{2}. \tag{6.14}$$

The energy related to the fluctuations of the phase is the difference between the change of energy due to these fluctuations without the gap and with the single particle gap. As the condensation energy is independent of the position of the gap in k space, fluctuations do not change this energy. The change of the energy for the configuration without the gap is given at zero temperature, using the linear dispersion relation $d\epsilon = \pm\hbar v_F\, dk$, by

$$E\left[\frac{d\phi}{dx}\right] = 4\int_0^{\phi'/2} \frac{\hbar v_F k dk}{2\pi} = \frac{\hbar v_F}{2\pi}\left(\frac{d\phi}{dx}\right)^2 \qquad (6.15)$$

where the factor of 4 accounts for the two sides of the Fermi surface and for the two spin directions. By virtue of $n(\epsilon_F) = N_e/\hbar k_F v_F = (\pi\hbar v_F)^{-1}$, Eq. (6.15) reduces to

$$E\left[\frac{d\phi}{dx}\right] = \frac{n(\epsilon_F)\hbar^2 v_F^2}{4}\left(\frac{d\phi}{dx}\right)^2. \qquad (6.16)$$

This, together with Eqs. (6.5) and (6.7), gives

$$c = \frac{\hbar^2 v_F^2}{4|\Delta|^2}. \qquad (6.17a)$$

In terms of the coherence length $\xi_0 = \hbar v_F/\pi|\Delta|$,

$$c = \frac{\pi^2}{16}\xi_0^2. \qquad (6.17b)$$

The consequences of the temporal variations of the phase can be described similarly. Assuming a translational motion of the undistorted condensate, the drift velocity v_d is given in terms of the time varying phase by

$$v_d = \frac{1}{2k_F}\frac{d\phi}{dt}. \qquad (6.18)$$

As a consequence, inversion symmetry in k space is broken, as shown in Fig. 6.2, and the gap, if tied to the moving condensate, appears at

$$\pm k_F + \delta k = \pm k_F + \frac{1}{2v_F}\frac{d\phi}{dt} \qquad (6.19)$$

as $\hbar\delta k = mv_d$. In the presence of a condensate moving with a velocity v_d the dispersion relation is derived as follows (Boriak and Overhauser, 1977; Schulz, 1977). We write the momentum in the moving reference frame as

$$\hbar\tilde{k} = \hbar k + mv_d$$

and then the dispersion relation in the moving reference frame is related to the dispersion relation in the laboratory frame as

$$\epsilon(k) = \tilde{\epsilon}(\tilde{k}) + m\tilde{v}v_d + \tfrac{1}{2}mv_d^2. \tag{6.20}$$

Following Eq. (3.34) the dispersion relation in the moving reference frame may be written as

$$\tilde{\epsilon}(\tilde{k}) = \pm\left(\hbar^2 v_F^2\left(\tilde{k} - \tilde{\epsilon}_F\right)^2 + \Delta^2\right)^{1/2} \tag{6.21}$$

and the dispersion relation in the laboratory frame is given (neglecting the last term in Eq. (6.20)) as

$$\epsilon(\tilde{k}) = \pm\left(\hbar^2 v_F^2\left(\tilde{k} - \tilde{k}_F\right)^2 + \Delta^2\right)^{1/2} \tag{6.22a}$$

$$+ \frac{mv_F v_d\left(\tilde{k} - \tilde{k}_F\right)}{\hbar^2 v_F^2\left(\tilde{k} - \tilde{k}_F\right)^2 + \Delta^2}$$

where we have used the relation $\tilde{v} = \partial\tilde{\epsilon}/\partial\tilde{k}$.

As $\tilde{k} - \tilde{k}_F = (k - k_F) + mv_d$, we obtain the following dispersion relation

$$\epsilon(k) = \pm\left(\hbar^2 v_F^2(k - k_F - mv_d)^2 + \Delta^2\right)^{1/2} \tag{6.22b}$$

$$\pm \frac{mv_F v_d(k - k_F - mv_d)}{\left(\hbar^2 v_F^2(k - k_F - mv_d)^2 + \Delta^2\right)^{1/2}}$$

which is shown in Fig. 6.2b.

It is easy to see that $\epsilon(\pm k_F)$ and therefore the energy gap remains unchanged, and all the modifications occur away from the (displaced) Fermi surface. The electric current density which is related to the displaced Fermi see,

$$j = \int_{k_F - mv_d}^{k_F + mv_d} \frac{\partial\epsilon}{\partial k}\, dk = \epsilon(k_F + mv_d) - \epsilon(k_F - mv_d) = 0. \tag{6.23}$$

The moving condensate, however leads to an electric current

$$j_{\mathrm{DW}} = N_e e v_d = -\frac{e}{\pi}\frac{d\phi}{dt} \tag{6.24}$$

as $N_e = 2k_F/\pi$.

The excess energy associated with the time dependent phase is evaluated using the same argument which has been used to evaluate the elastic energy. In the absence of the gap the change of the energy is given by

$$E\left[\frac{d\phi}{dt}\right] = 4\int_0^{k_F} \frac{\dot{\phi}}{2v_F} v_F k dk/2 = \frac{n(\epsilon_F)}{4}\left(\frac{d\phi}{dt}\right)^2 \tag{6.25}$$

The kinetic energy term as given by Eq. (6.21) represents the change of the total electronic energy and does not include the kinetic energy of the ions which exhibit an oscillating motion for a time-dependent phase. This energy leads to an enhancement of the dynamical mass, which can be estimated as follows: for a charge density wave moving with a uniform drift velocity v_d the total energy is given by

$$E_k = \frac{1}{2}mv_d^2 + \frac{1}{2}M(\omega/\Delta u)^2, \tag{6.26}$$

where M is the ionic mass, Δu is the atomic displacement, and ω is the angular frequency of the ionic oscillations. Here ω and v_d are related through $\omega = 2\pi v_d/\lambda_0$, and consequently

$$E\left[\frac{d\phi}{dt}\right] = \frac{1}{2}mv_d^2\left(1 + \frac{4\pi^2/m^2}{\lambda_0^2}\frac{M}{m^2}\right) = \frac{1}{2}m^* v_d^2, \tag{6.27}$$

where m^* is called the effective mass. With $\lambda_0 = \pi/k_F$, and $\Delta u = \Delta/g\omega_{2k_F}$ we obtain

$$\frac{m^*}{m} = 1 + \frac{4k_F^2 n(\epsilon_F)\Delta^2}{N_e g \omega_{2k_F}^2 m} \tag{6.28}$$

which, with $k_F^2 = 2m\epsilon_F/\hbar^2$ and $N_e = 2n(\epsilon_F)\epsilon_F$ (both appropriate for a one-dimensional band) and with $\lambda = gn(\epsilon_F)$ leads to

$$\frac{m^*}{m} = 1 + \frac{4\Delta^2}{\lambda\hbar^2\omega_{2k_l}^2}. \tag{6.29}$$

The total kinetic energy is then given by

$$E\left[\frac{d\phi}{dt}\right]=\frac{n(\epsilon_F)}{4}\frac{m^*}{m}\left(\frac{d\phi}{dt}\right)^2 \tag{6.30}$$

and consequently, using Eq. (6.12),

$$d=\frac{1}{4}\frac{m^*}{m|\Delta|^2}. \tag{6.31}$$

The arguments which lead to Eq. (6.25) apply in the $\omega \to 0$, $q \to 0$ limit. At finite frequencies and wavevector (Virosztek and Maki, 1993)

$$\frac{m^*}{m}=\frac{m^*(\omega=0,q=0)}{m}\left[1+\frac{2}{3}\left(\frac{\omega}{2\Delta}\right)^2-\frac{2}{3}\left(\frac{v_Fq}{2\Delta}\right)^2\right]^{-1} \tag{6.32}$$

at zero temperature, the expression is appropriate in the $\omega, v_Fq << \Delta$ limit. This frequency and wavevector dependence then modifies the dispersion relation of the amplitude and phase mode.

By inserting the appropriate expressions for the Ginzburg-Landau parameters into Eq. (6.11) the dispersion relation for the phase excitations is given at zero temperature by

$$\omega_\phi=\left(\frac{m}{m^*}\right)^{1/2}v_Fq=c_\phi q \tag{6.33}$$

where c_ϕ is called the phason velocity. The amplitude mode frequency, with $a=\frac{1}{2}$ and Eq. (6.11a), is

$$\omega_A=\lambda^{1/2}\omega_{2k_f} \tag{6.34}$$

at $q=0$. This is the same as Eq. (6.4). Eq. (6.11b) together with the wavevector-dependent effective mass, given by Eq. (6.32) gives (Lee, Rice, and Anderson, 1979)

$$\omega_A=\left(\lambda\omega_{2k_F}^2+\frac{1}{3}\frac{m}{m^*}v_F^2q^2\right)^{1/2} \tag{6.35}$$

for long wavelength fluctuations. The dispersion relations of the

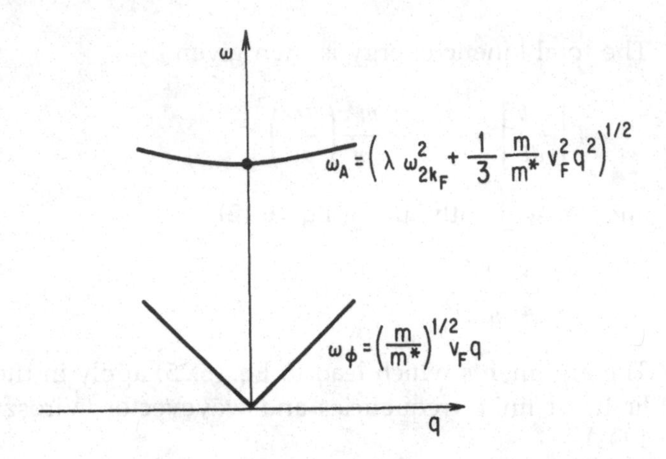

Figure 6.3. The dispersion relation of the phase and amplitude mode near $q = 0$ for a charge density wave ground state.

phase and amplitude modes are displayed in Fig. 6.3, and the two modes are sketched in Fig. 6.4 for a finite wavevector q.

The above arguments apply at zero temperature; the temperature dependence of the Ginzburg-Landau parameters a, c, and d are determined by the temperature dependence of the gap $|\Delta(T)|$ and of the condensate density $f_0(T)$. With these included, the free energy reads (Fukuyama and Lee, 1978; Brazoskii and

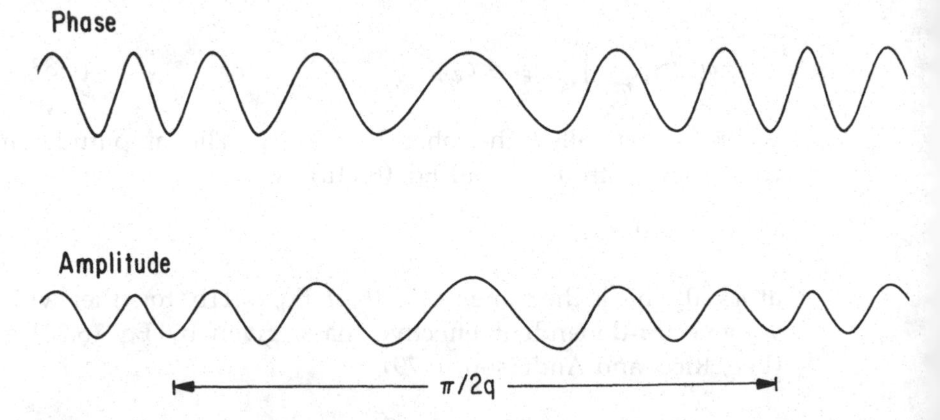

Figure 6.4. Phase and amplitude excitations with a finite wavevector q. Only the charge density is indicated on the figure, there are corresponding modifications in the atomic positions.

Dzyaloshimskii, 1976)

$$F = F(0) + \frac{1}{4} n(\epsilon_F) f(T) \int dx \tag{6.36}$$

$$\left\{ -2|\Delta|^2 + 2\delta^2 + |\Delta|^{-2} \left[\hbar^2 v_F^2 \left(\frac{\partial \delta}{dx} \right)^2 + \frac{m^*}{m} \left(\frac{\partial \delta}{\partial t} \right)^2 \right] \right.$$

$$\left. + \hbar^2 v_F^2 \left(\frac{\partial \phi}{\partial x} \right)^2 + \frac{m^*}{m} \left(\frac{\partial \phi}{\partial t} \right)^2 \right\}.$$

At finite temperatures both ω_A and c_ϕ are determined by the temperature dependence of $\Delta(T)$ and $f(T)$. The former is determined by Eq. (3.55) and the latter by Eq. (3.58). For the frequency of the phase excitations for small wavelengths ($\omega_A \gg v_F q$) the dynamic limit applies, and

$$\omega_A = \lambda^{1/2} \omega_{2k_F} [f_d(t)]^{1/2} \tag{6.37}$$

reduces to

$$\omega_A \sim \frac{1.55}{4} \lambda^{1/2} \omega_{2k_F} \left(1 - \frac{T}{T_{\mathrm{CDW}}} \right)^{1/4} \tag{6.38}$$

close to the transition temperature T_{CDW}. The temperature dependence of the phase and amplitude mode are displayed in Fig. 6.5.

The temperature dependence of the effective mass, which determines the temperature dependence of the phase velocity,

$$\frac{m^*(T)}{m} = 1 + \frac{4\Delta^2(T)}{\hbar^2 \lambda \omega_{2k_F}^2} \frac{1}{f_s(T)} \tag{6.39}$$

is also determined both by the temperature dependent single particle gap and by the temperature dependent condensate density as $c_\phi < v_F$ the static limit is appropriate. By inserting the appropriate expressions for $\Delta(T)$ and $f_s(T)$, we obtain

$$\frac{m^*(T)}{m^*(T=0)} = 1.51 \tag{6.40}$$

close to the transition temperature. The frequency of the amplitude fluctuations approaches zero as $T \to T_{\mathrm{CDW}}^{\mathrm{MF}}$, while the velocity of the phase fluctuations remains finite as $T \to T_{\mathrm{CDW}}^{\mathrm{MF}}$.

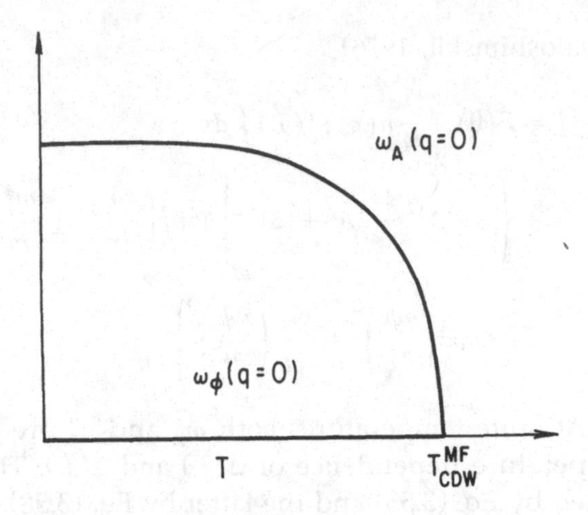

Figure 6.5. Temperature dependence of the frequency of the phase and amplitude modes, $\omega_A(q=0)$ and $\omega_\phi(q=0)$.

The above analysis applies in the absence of long-range Coulomb forces. Consequently the amplitude and phase mode dispersion relations are appropriate only for transverse modes, which do not involve the redistribution of electronic charges due to fluctuations, or for longitudinal modes where Coulomb interactions are screened by uncondensed electrons; i.e., at temperatures close to $T_{\mathrm{CDW}}^{\mathrm{MF}}$.

At zero temperature, with no screening by quasiparticles, the long range Coulomb interactions become important, and this leads to a gap in the phason spectrum for longitudinal excitations as first pointed out by Anderson and Higgs (Anderson, 1984). The argument which leads to the gap frequency is analogous to the argument which leads to the plasma frequency in metals. A density wave displacement Δu relative to the underlying ionic background leads to a polarization

$$\Delta P_0 = ne\Delta u \tag{6.41}$$

which gives rise to an electric field

$$E = -\frac{4\pi P}{\epsilon_0} \tag{6.42}$$

where ϵ_0 is the background dielectric constant due to single particle excitations across the semiconducting gap. The equation of motion for the condensate is then

$$m^* \left(\frac{dx}{dt} \right)^2 = -4\pi ne^2 \epsilon_0^{-1} x \tag{6.43}$$

and this defines a characteristic frequency

$$\omega_p^* = \left(\frac{m}{m^*} \right)^{1/2} \omega_p \epsilon_0^{-1/2} \tag{6.44}$$

where

$$\omega_p = \left(\frac{4\pi ne^2}{m_b} \right)^{1/2} \tag{6.45}$$

is the plasma frequency of the metallic state above T_{CDW}. The dielectric constant is given by

$$\epsilon_0 = 1 + \frac{1}{6} \left(\frac{\omega_p}{\Delta} \right)^2 \tag{6.46}$$

for a one-dimensional semiconductor with a gap Δ, and consequently

$$\omega_\phi \simeq 6^{-1/2} \left(\frac{m}{m^*} \right)^{1/2} |\Delta|^2 = \left(\frac{3\lambda}{2} \right)^{1/2} \omega_{2k_F} \tag{6.47}$$

by virtue of Eq. (6.30).

In the case of screening by the thermally excited quasiparticles the problem becomes somewhat complicated, and the frequency and damping constant associated with the phase excitations depend on factors like the Thomas-Fermi wavevector q_0, the diffusion content D_{qp} of the quasiparticles, and whether the static or dynamic limit ($\omega \gtrless v_F q$) applies. Broadly speaking fluctuations with $q < q_0$ are screened, while no screening occurs for $q > q_0$. The Thomas-Fermi wavevector has been evaluated (Maki and Virosztek, 1992), and its temperature dependence is given by

$$q_0 \simeq \omega_p v_F^{-1} (1 - f_s)^{1/2}. \tag{6.48}$$

where ω_p is the plasma frequency and f_s is the static condensate density. With such screening included, the following dispersion relation is obtained for the longitudinal phase mode (Lee and

Fukayama, 1978; Kurihara, 1980; Barisic, 1985; Nakane and Takada, 1987)

$$\omega_\phi^2 = \left\{ c_0^2 + \frac{m}{m^*} \frac{\Delta^2}{q^2 + q_0^2} \right\} q^2 \tag{6.49}$$

which for the two limits reduces (in the static limit) to

$$\omega_\phi^2 = \frac{c_0^2 q^2}{1 - f_s}, \qquad q \ll q_0 \tag{6.50a}$$

$$\omega_\phi^2 = \frac{3}{2} \lambda \omega_{2k_F}^2 f_s + c_0^2 q^2 \qquad q \gg q_0 \tag{6.50b}$$

with $c_0 = (m/m^*)^{1/2} v_F$. The dispersion relation given by the above equations is sketched in Fig. 6.6 at a temperature somewhat below T_{CDW}. Due to quasiparticle damping, the mode also becomes diffusive near $q = 0$ and strongly damped for finite q values in the region near q_0 and below.

The situation is analogous to that which occurs for superconductors, and to that known as the Carlson-Goldman mode (1973, 1975). At long wavelengths the condensate current is compensated by the quasiparticle current and the mode remains acoustic;

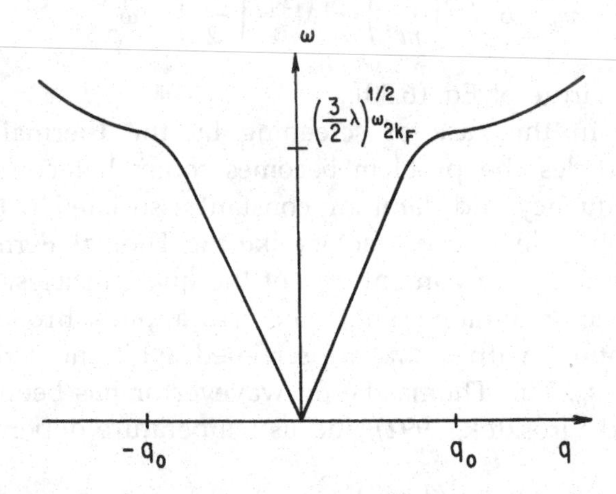

Figure 6.6. The dispersion relations of the phase mode at finite temperature in the presence of Coulomb effects and quasiparticle screening. The screening wavevector q_0 is defined in the text. The mode becomes diffusive near $q = 0$, and is strongly damped for $q < q_0$.

for $q > q_0$ the condensate and quasiparticle currents oscillate in phase and are therefore subjected to Coulomb forces. As the quasiparticle density is strongly temperature dependent, ω_ϕ also has a strong temperature dependence (Nakane and Takada, 1985; Maki and Virosztek, 1990). Obviously, Coulomb effects do not play a role for the transverse modes; they remain acoustic in the entire temperature range. Similarly the amplitude mode remains unaffected by long range Coulomb effects.

In three dimensions, the fluctuations of all three spatial coordinates contribute to the free energy and the elastic term reads

$$F_{el} = \frac{1}{4} n(\epsilon_F)\hbar^2|\Delta|^{-2}\left[\int v_{F_x}^2\left(\frac{d\Delta}{dx}\right)^2 dx + \int v_{F_y}^2\left(\frac{d\Delta}{dy}\right)^2 dy + \int v_{F_z}^2\left(\frac{d\Delta}{dz}\right)^2 dz\right] \tag{6.51}$$

with the on-chain Fermi velocity $v_{F_x} \gg v_{F_y}, v_{F_z}$. With the coherence length $\xi = \hbar v_F/\pi\Delta$ proportional to the Fermi velocity, Eq. (6.51) is usually written as

$$F = \frac{1}{4}\frac{n(\epsilon_F)\hbar^2 v_F}{a_y a_z|\Delta|^2}\int d\vec{r}\left[\left(\frac{\partial\Delta}{\partial x}\right)^2 + \left(\frac{\xi_y}{\xi_\parallel}\right)^2\left(\frac{\partial\Delta}{\partial y}\right)^2 + \left(\frac{\xi_z}{\xi_\parallel}\right)^2\left(\frac{\partial\Delta}{\partial z}\right)^2\right]. \tag{6.52}$$

Where v_F refers to the Fermi velocity along the chain direction. This leads to different dispersion relations for CDW fluctuations in the different directions, and

$$c_{\phi x} = \left(\frac{m}{m^*}\right)^{1/2} v_{F_x} q \qquad \text{parallel}$$
$$c_{\phi y}, c_{\phi z} = \left(\frac{m}{m^*}\right)^{1/2} v_{F_{y,z}} q \quad \text{perpendicular.} \tag{6.53}$$

The three amplitude modes have the same frequency at $q = 0$, but because of the different Fermi velocities, they display different q dependences for oscillations involving amplitude modes in the three different directions, as implied by Eq. (6.36).

The excitations of the amplitude and phase also contribute to the various thermodynamic quantities. At low temperatures, because of the gap in the excitation spectrum, the contribution from amplitude excitations is small. The excitations of the phase can be described using the formalism applied to acoustic phonons. We write the Hamiltonian density $\mathscr{H}$, corresponding to the free energy Eq. (6.5), as

$$\mathscr{H} = \frac{\pi^2}{2\rho} + c\left(\frac{d\phi}{dx}\right)^2 \tag{6.54}$$

where π is the momentum density. The mass density ρ is given by

$$\rho = \left[\frac{n(\epsilon_F)}{2}\frac{m^*}{m}\right]^{-1}. \tag{6.55}$$

We express $\phi(x)$ and $\pi(x)$ in terms of the variables Q'_k and P'_k as

$$\phi(x) = L^{-1/2}\sum_q Q'_q e^{iqx} \tag{6.56a}$$

$$\pi(x) = L^{-1}\sum_q P'_q e^{iqx} \tag{6.56b}$$

by using periodic boundary conditions over the length L. Then the Hamiltonian becomes

$$\mathscr{H} = \int \mathscr{H}\, dx = \sum_q \left(\frac{1}{2\rho}P'_q P'_{-q} + cq^2 Q'_q Q'_{-q}\right) \tag{6.57}$$

which, by using the transformation

$$a_q^\dagger = -i(2\rho\omega_\phi)^{-1/2} P'_q + \left(\frac{c}{\omega_\phi}\right)^{1/2} qQ'_{-q} \tag{6.58a}$$

$$a_q = i(2\rho\omega_\phi)^{-1/2} P'_q + \left(\frac{c}{\omega_\phi}\right)^{1/2} qQ'_q, \tag{6.58b}$$

becomes

$$\mathscr{H} = \sum_q \omega_\phi a_q^\dagger a_q. \tag{6.59}$$

The dispersion relation, Eq. (6.34), is given in terms of c and the mass density as $\omega_\phi = (2c/\rho)^{1/2}q$, which, as has been shown earlier, is $\omega_\phi = c_\phi q$ where c_ϕ is the phase velocity.

The calculation of the low temperature specific heat follows the calculation of the specific heat due to phonons, and in the Debye approximation we obtain

$$C_V^\phi \simeq (2nk_B)^3 \left(\frac{T}{\Theta_D}\right)^3 \tag{6.60}$$

where n is the total number of modes per unit length; for a CDW period of λ_0 this is given by $n = \lambda_0^{-1}$. The cutoff wavevector $q_{\max}$ is given by the relation

$$n = \frac{4\pi}{3}\left(\frac{1}{2\pi}\right)^3 q_{\max}^3 \frac{T_{\mathrm{CDW}}}{T_F} \tag{6.61}$$

where the factor T_{CDW}/T_F is due to the fact that a significant fraction of the lattice entropy has been removed at the phase transition. The Debye temperature then is

$$k_D \theta_D^\phi = q_{\max} c_\phi = 2\pi \left(\frac{3n}{4\pi}\right)^{1/3} v_F \left(\frac{m}{m^*}\right)^{1/2} \left(\frac{T_F}{T_{\mathrm{CDW}}}\right)^{1/3}. \tag{6.62}$$

In case of isotropic phason and phonon dispersion relations, the relative contributions of the two types of excitation to the specific heat is given by

$$\frac{C_V^\phi}{C_V^{\mathrm{ph}}} = \left(\frac{n_\phi}{n_{\mathrm{ph}}}\right)^3 \left(\frac{\theta_D^{\mathrm{ph}}}{\theta_D^\phi}\right)^3. \tag{6.63}$$

The number of phonon modes per unit volume is proportional to a^{-3} with a the lattice constant, and therefore $n_\phi/n_{\mathrm{ph}} = (a/\lambda)^3$. With a typical sound velocity $c_0 \sim 5 \times 10^5$ cm/sec, and a typical phase velocity $c_\phi = v_F(m/m^*)^{1/2} = 10^5$ cm/sec, the two contributions are of the same order of magnitude.

The above arguments apply for an isotropic phase excitation spectrum. In the case of an anisotropic dispersion relation such as that given by Eq. (6.52), the specific heat is modified and is given by

$$c_V^{\phi} = c_{V(\text{isotropic})}^{\phi} \cdot \left(\frac{\xi_{\parallel}^2}{\xi_y \xi_z} \right) \tag{6.64}$$

at low temperatures. As $\xi_y, \xi_z \ll \xi_{\parallel}$, a significant enhancement of the specific heat is expected due to anisotropy effects at low temperatures.

6.2 Excitations of the Spin Density Wave Ground State

Because of the magnetic character of the SDW ground state, the excitations are different from the excitations which occur for the CDW ground state. With both spin and charge degrees of freedom available, various types of amplitude and phase excitations may occur.

We write the time dependence of the phase fluctuations in the two subbands as $\delta\phi_{\uparrow} = e^{i\omega t}$ and $\delta\phi_{\downarrow} = e^{i\omega t + \phi}$. Similarly, the amplitude fluctuations are given by $\delta A_{\uparrow} = e^{i\omega t}$ and $\delta A_{\downarrow} = e^{i\omega t + \phi}$. An arbitrary phase ϕ leads, in both cases, to a periodic modulation of the charge density, such excitations however, are energetically not favored. To see what happens, consider the cases for which $\phi = 0$ or $\phi = \pi/2$; the various possibilities are shown in Fig. 6.7 in the $q = 0$ limit.

The top part of the figure indicates the displacement of both CDW modulations in the same direction ($\phi = 0$) and this leads in turn to the displacement of the spin density wave modulation without leading to an oscillating charge density wave component. The excitation is analogous to the phase excitations discussed for charge density waves. The next two parts of the figure describe two different excitations, both related to the change of the amplitudes of the density wave modulations in the two subbands. The first represents the amplitude fluctuations for which $\phi = \pi$. This however should lead to a charge density wave component of the order parameter and is consequently not allowed. The amplitude fluctuations may also occur in phase, and this leads to no charge redistribution but a change of the amplitude of the spin density wave modulation. This excitation corresponds to the amplitude mode as discussed earlier. The fourth possibility—charge density wave modulations, oscillating in opposite phase with $\phi =$

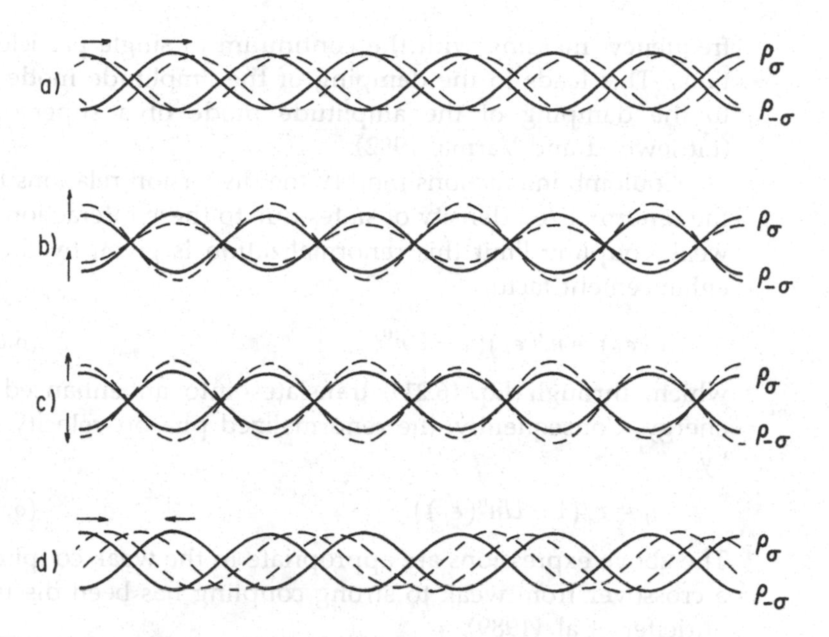

Figure 6.7. The amplitude and phase excitations of the spin density wave state in the $q = 0$ limit. The full lines represent the charge density wave modulation in the two subbands; the dashed lines represent the various amplitude and phase modes.

π—would also lead to an oscillating charge density wave component, and is thus forbidden.

This picture which follows from the above considerations for the charge excitations is similar to that described for charge density waves, with some modifications. The phase and amplitude excitations are analogous to those given by Eqs. (6.34) and (6.35) with two important differences. First, because the SDW modulation is not coupled to the lattice, the effective mass is the same as the bandmass, $m^* = m_b$, and consequently

$$\omega_\phi = v_F q. \tag{6.65}$$

Second, instead of Eq. (6.35), the frequency of the amplitude mode at $q = 0$ is given using $a = \frac{1}{2}$ and $d = 1/4\Delta^2$ (Psaltakis, 1984), by

$$\omega_A = \left((2\Delta/\hbar)^2 + (v_F q)^2\right)^{1/2} \tag{6.66}$$

The amplitude mode therefore is expected to occur above the gap frequency, merging with the continuum of single particle excitations. This leads to the damping of the amplitude mode, similar to the damping of the amplitude mode of a superconductor (Littlewood and Varma, 1982).

Coulomb interactions modify the dispersion relations through the renormalized density of states due to these interactions. In the weak coupling limit this renormalization is given by the Stoner enhancement factor

$$n(\epsilon_F) = n^0(\epsilon_F)(1 - Un^0(\epsilon_F))^{-1} \tag{6.67}$$

which, through Eq. (6.21), translates into an enhanced kinetic energy. Consequently, the renormalized phason velocity is given by

$$v_F = v_F^0(1 + Un^0(\epsilon_F)). \tag{6.68}$$

The above expressions are appropriate in the weak coupling limit; a crossover from weak to strong coupling has been discussed by Shrieffer et al. (1989).

Next we turn to spin excitations. In the absence of spin-orbit and dipole-dipole interactions, the spin degrees of freedom have full rotational symmetry, this leads to excitations which are those of a 1D Heisenberg antiferromagnet. These excitations are described by the Hamiltonian in Eq. (4.35) which accounts for the static magnetic properties of the SDW ground state. Consequently, results which have been obtained for antiferromagnets can be adapted to describe the magnetic excitations of the spin density wave ground state.

In the absence of magnetic anisotropy fields, the equation of motion

$$\hbar \frac{d}{dt}\langle \vec{S} \rangle = u\left[\langle \vec{S} \rangle \times \vec{H}_{\text{eff}}\right] \tag{6.69}$$

leads to the dispersion relation (Kittel, 1963) for the two magnon branches

$$\omega_s = 2J_{\text{eff}}aq = 2J_{\text{eff}}\frac{\lambda_0}{2}q \tag{6.70}$$

The effective coupling constant J_{eff} is given in terms of the parameter s of the metallic state by Eq. (4.38). This then leads to the dispersion relation

$$\omega_s = v_F q. \tag{6.71}$$

Coulomb interactions and an enhanced density of states as given by Eq. (6.56) lead, through $n(\epsilon_F) = (\pi\hbar v_F)^{-1}$, to a decrease of the Fermi velocity and therefore the dispersion relation reads as

$$\omega_s = v_F^0(1 - Un^0(\epsilon_F))q \tag{6.72}$$

in the case of Coulomb interactions.

Spin orbit and dipole-dipole interactions remove the rotational symmetry of the magnetic excitations and lead to an energy gap in the spin wave excitation spectrum. The gap and the dispersion relation depend sensitively on the orientation of the external magnetic field with respect to the preferred axis of magnetization. For anisotropy constants D^* and E^* the spin wave spectrum has two branches, and for zero magnetic field, in the $q = 0$ limit

$$\hbar^2\omega_+^2 = \frac{\mu}{\mu_B}(D^* + E^*) J_{\text{eff}} \tag{6.73}$$

$$\hbar^2\omega_-^2 = \frac{\mu}{\mu_B}2E^* J_{\text{eff}}. \tag{6.74}$$

The magnetic field dependence of the resonant frequencies can also be calculated for various directions of the magnetic field (Torrance et al., 1982).

6.3 Experiments on Charge Density Waves: Neutron and Raman Scattering

Both the phase and amplitude mode dispersion relations can be examined using inelastic neutron scattering. Phase excitations lead to the displacement of the electronic charge density, and consequently the phase mode is optically active. The electrodynamics of the ground state will be discussed in Chapter 9. No charge fluctuation occurs for amplitude fluctuations, therefore the amplitude mode is Raman active. Both types of optical experiments have been extensively used to explore the excitations of the CDW ground state.

The phase mode has been measured in $K_{0.3}MoO_3$ by inelastic neutron scattering at temperatures somewhat below T_{CDW} (Pouget et al., 1991; Escribe-Filippini et al., 1987). The constant frequency scans together with the acoustic mode dispersion relation are displayed in Fig. 6.8. The linear dispersion observed is in full agreement with Eq. (6.7) and gives a phase velocity of $c_\phi =$

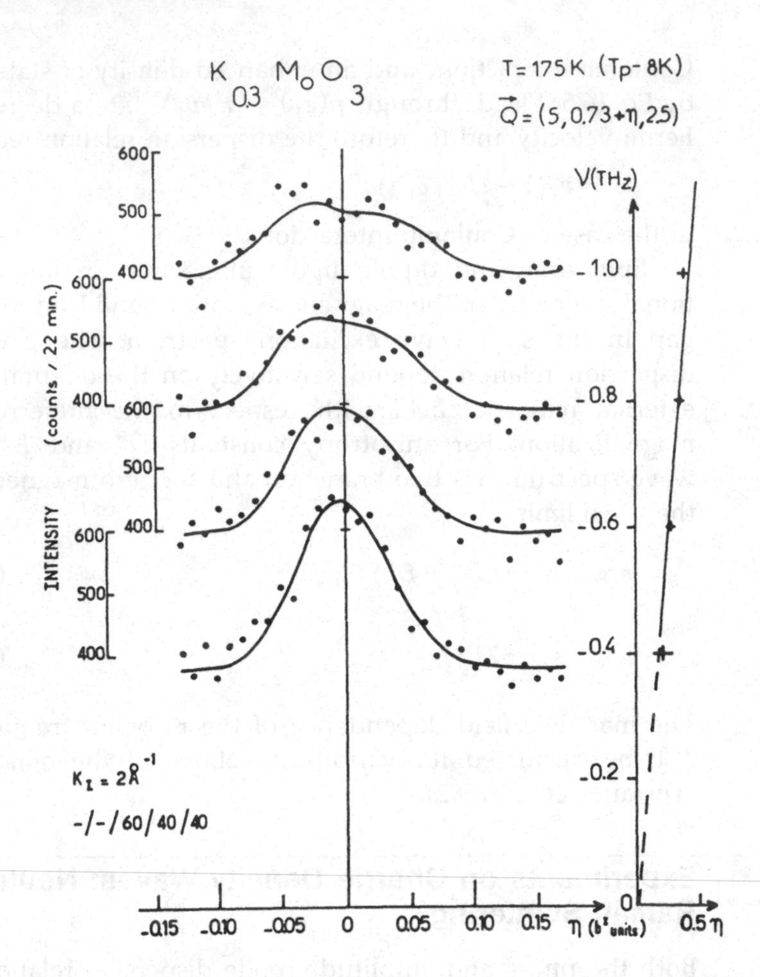

Figure 6.8. Constant frequency scans (left side) and dispersion (right side) of the phase mode branch of $K_{0.3}MoO_3$ and $T = 175$ K. The slope gives a phason velocity $v_\phi = 3.3 \pm 0.5 \times 10^5$ cm/sec in the chain direction (Escribe-Filippini et al., 1987).

3.3×10^5 cm/sec. This, together with the Fermi velocity as given in Table 2.1 results in a mass enhancement $m^*/m \simeq 100$.

The mass enhancement can also be calculated using Eq. (6.25). The single particle gap $2\Delta = 1400$ K, together with λ and ω_{2k_F} gives $m^*/m = 600$, in good agreement with the results obtained from the measurement of the phase velocity. The temperature dependence of c_ϕ has also been examined (Hennion et al., 1992) and was found to increase with decreasing temperature in accordance with what is expected from the increased role of Coulomb

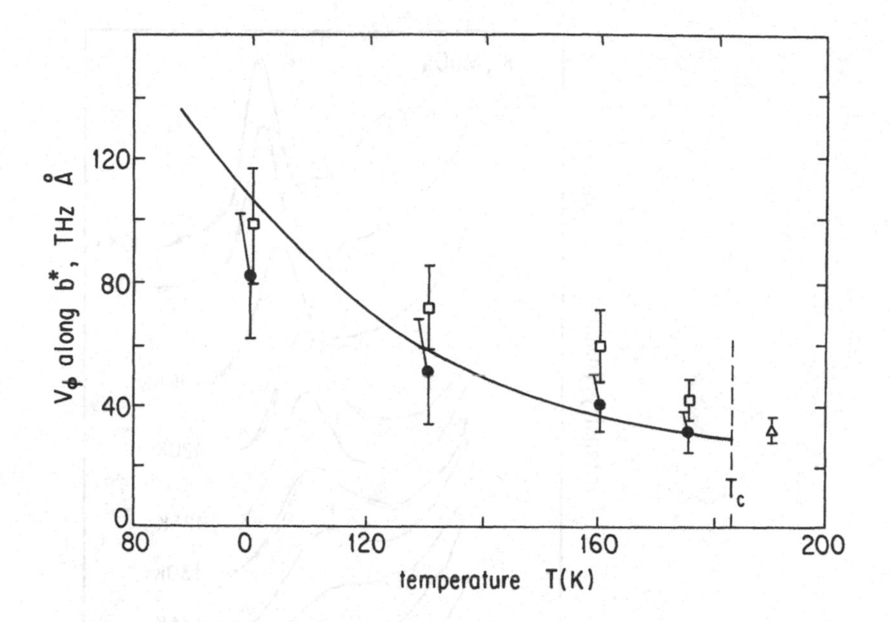

Figure 6.9. Temperature dependence of the phason velocity measured in $K_{0.3}MoO_3$ by neutron scattering (Hennion et al., 1992). The full line is the temperature dependence calculated by taking Coulomb effects into account (Maki and Virosztek, 1992; Maki and Grüner, 1991).

interactions as the thermally excited electrons are progressively removed. Fig. 6.9 shows c_ϕ, measured at different temperatures. The full line is calculated by evaluating the temperature dependence of the phase velocity due to Coulomb effects at finite temperatures (Maki and Grüner, 1991; Maki and Virosztek, 1992).

The phason velocity was also found to be anisotropic, with the velocities measured perpendicular to the chain direction significantly smaller than the phason velocity along the chains. This anisotropy is in broad agreement with the anisotropic band structure of $K_{0.3}MoO_3$ and with the anisotropy of the coherence length. One also finds that the frequency of the mode remains finite as $q \to 0$, and this gap can be explained as the pinning of the phase mode by impurities; such pinning acts as a restoring force, shifting the phase oscillation frequency to a finite value as $q \to 0$.

The amplitude mode has been measured in several compounds both by neutron and by Raman scattering (Travaglini et al., 1983). The Raman spectrum, obtained at various temperatures on $K_{0.3}MoO_3$ is shown in Fig. 6.10 (Travaglini et al., 1983).

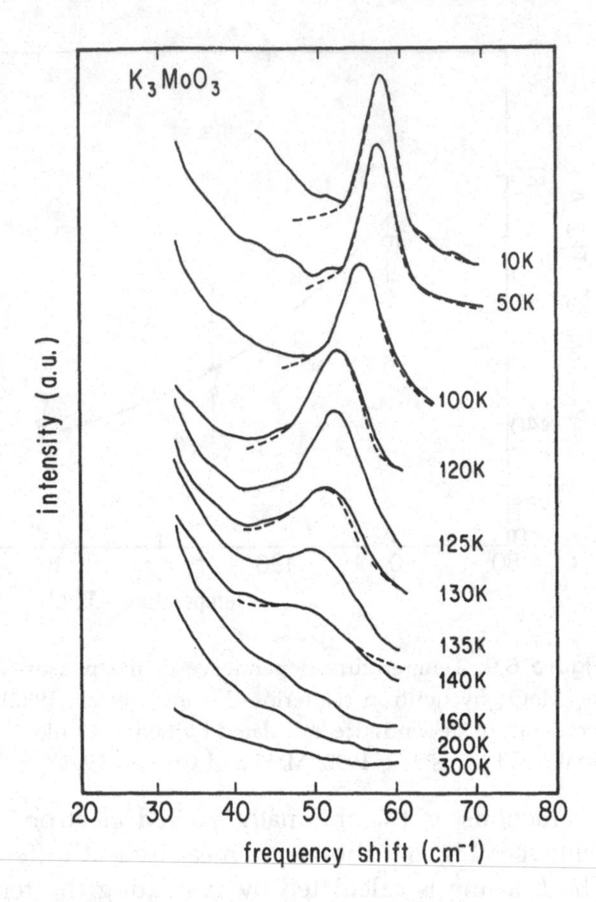

Figure 6.10. Raman spectrum of $K_{0.3}MoO_3$ in the charge density wave state (Travaglini et al., 1983).

The mode, is analyzed in terms of a harmonic oscillator fit

$$I(\omega) = \frac{\beta(\omega) - 1}{\left(\omega_A^2 - \omega^2\right)^2 - (2\omega\Gamma)^2} \tag{6.75}$$

where ω_A is the oscillator frequency, Γ is the damping constant, and $\beta(\omega)$, is the Bose-Einstein factor. Fits to Eq. (6.75) are also displayed in the figure. As expected the intensity of the mode approaches zero at the phase transition, and Γ increases as the temperature increases.

The frequency of the amplitude mode, $\omega_A \simeq 1.7$ THz at $T \to 0$, is comparable to the frequency of the unrenormalized phonon

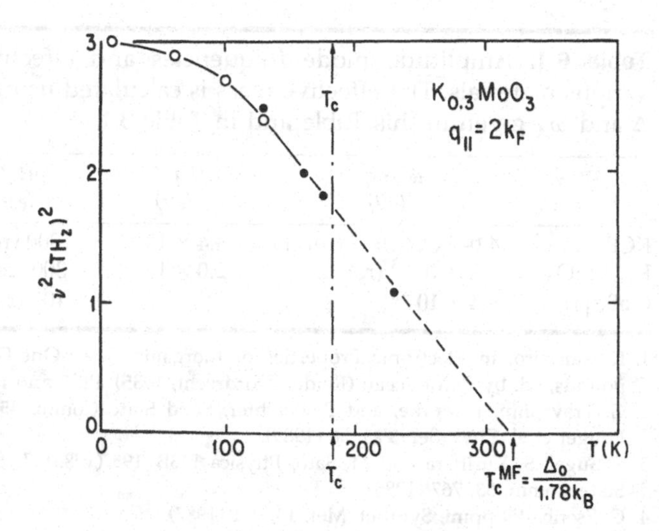

Figure 6.11. Temperature dependence of the amplitude mode frequency in $K_{0.3}MoO_3$. The dotted line is an extrapolation, assuming $\omega_a^2 \approx (T_{\text{CDW}}^{\text{MF}} - T)$ (Travaglini et al., 1983; Pouget and Comes, 1989).

mode $\omega_{2k_F} = 1.5$ THz. This then, through $\omega = \lambda^{1/2}\omega_{2k_F}$ would imply that λ is close to number 1 in broad agreement with the d value arrived at in Chapter 3. The frequency decreases with increasing temperature, in qualitative agreement with what follows from the temperature dependent gap $\Delta(T)$ and condensate density $f(T)$. The temperature dependence $\omega_A(q = 0, T)$ is shown in Fig. 6.11 where measurements using both neutron (Pouget et al., 1991) and Raman scattering (Travaglini et al., 1983) are included. The full line is a fit to Eq. (6.38) with $T_{\text{CDW}}^{\text{MF}} = 320$ K, well above the transition temperature of $T_{3\text{D}} = 180$ K. This, as discussed in the previous chapter, is due to fluctuation effects which reduce the transition where 3D order develops with respect to the mean field transition temperature.

The amplitude mode frequencies at $T = 0$ measured in various compounds with a charge density wave ground state are collected in Table 6.1; together with the values calculated using Eq. (6.35), the electron-phonon coupling constants λ, and the unrenormalized phonon frequencies ω_{2k_F} as given in Tables 2.1 and 3.1. The effective mass values, measured by using neutron scattering or optical methods, together with the calculated values using Eq. (6.25), are also displayed.

Table 6.1. Amplitude mode frequencies and effective masses for various materials. The effective mass is calculated using the values of Δ and ω_A given in this Table and in Table 3.1

	ω_A(exp) (eV)	ω_A(Eq. 6.35) (eV)	m^*/m (exp)	m^*/m (Eq. 6.29)
KCP	$4.0\text{–}5.0 \times 10^{-3}$ (ref. 1)	4.4×10^{-3}	1000 (ref. 1)	660
$K_{0.3}MoO_3$	5.2×10^{-3} (ref. 2)	2.0×10^{-3}	300 (ref. 4)	260
$(TaSe_4)_2I$	9.4×10^{-3}		10^4 (ref. 5)	

1. K. Carneiro, in Electronic Properties of Inorganic Quasi-One-Dimensional Compounds, ed. by P. Monceau (Reidel, Dordrecht, 1985), Pt. 2 and references therein.
2. G. Travaglini, I. Morke, and P. Wachter, Solid State Comm. **45**, 289 (1983). J. P. Pouget et al. Phys. Ser. **T25**, 58 (1989).
3. S. Sugai, S. Kurihara and M. Sato, Physica 143B, 195 (1986), T. Sekine et al. Solid State Comm. **53**, 767 (1985).
4. C. Escribe-Filippini, Synthet. Met. **19**, 931 (1987).
5. T. W. Kim et al. Phys. Rev. **B43**, 6315 (1991).

It is expected that these excitations are accessible by thermal fluctuations and contribute to the various thermodynamic quantities. Because of the gap in the dispersion relation, contributions to the entropy from amplitude fluctuations are likely to be small. As discussed earlier, due to the linear dispersion relation, phase fluctuations lead to a specific heat which is proportional to T^3 at low temperatures where these fluctuations have a three-dimensional character. The strong anisotropy, however, may lead to a crossover to temperature regions where the character of these fluctuations is one- or two-dimensional. This in turn leads to temperature dependences different from the 3D case. A detailed analysis of the measured specific heat in various compounds has not been performed to date.

6.4 Experiments on Spin Density Waves: AFMR and Magnetization

The phase and amplitude modes which are related to charge excitations of the SDW state have not been measured, and neutron or Raman scattering studies of these materials have not been performed to date. Optical conductivity measurements, which couple to $q = 0$ phase excitations, are discussed in Chapter 9.

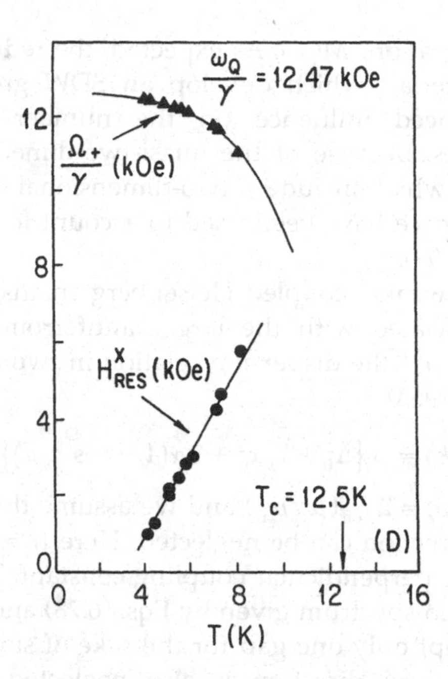

Figure 6.12. Observed resonance fields (circles) vs. temperature in $(TMTSF)_2PF_6$ for H along the hard axis, which give the temperature dependence of ω_+, see Eq. (6.73) (Torrance et al., 1982).

The spin excitations have been examined in various materials through measurements of antiferromagnetic resonance. These experiments add little to our understanding of the SDW state, but give clear evidence that Eq. (4.35) is an appropriate starting point in describing the magnetic properties of the ground state. The experiments, in general, are conducted at a fixed measurement frequency, while the magnetic field and/or temperature is varied. The resonance frequencies measured as a function of temperature are shown in Fig. 6.12. A comparison with calculated values gives a rough estimate of $\mu/\mu_B = 0.1$, in agreement with NMR and μSR studies (Torrance et al., 1982) discussed in Chapter 4.

As for the case of antiferromagnets, the thermally induced magnetic excitations also determine the temperature dependence of the magnetization. The magnetic susceptibility and antiferromagnetic resonance experiments can consistently be described in terms of an effective Heisenberg Hamiltonian, Eq. (4.35), which can also be used to evaluate the temperature dependence of the magnetization, $M(T)$. As expected, the reduced dimensionality of the materials which develop an SDW ground state also has a pronounced influence on the number of thermally excited magnons. Because of the quasi-two-dimensional band structure, models which include a two-dimensional dispersion relation and phase space have been used to account for experiments (Uemura et al., 1992).

For weakly coupled Heisenberg chains, keeping only the gap Δ_0 associated with the larger antiferromagnetic resonance frequency ω_+, the dispersion relation in two dimensions is given by (Kittel, 1963)

$$\omega(k) = \omega_e\{\Delta_0^2 + k_x^2 x^2 + 2\alpha(1 - \cos k_y y)\}^{1/2} \tag{6.76}$$

where $\omega_e = 2J_{\text{eff}}(\mu/\mu_B)$ and we assume that the dispersion in the third direction can be neglected. Here $\alpha = J_{\text{eff}}/J_\perp$,where $J_\perp$ is an effective perpendicular coupling constant. The gap in the magnon excitation spectrum given by Eqs. (6.73) and (6.74) is $\omega_e\Delta_0$ and we have kept only one gap for the sake of simplicity. The dispersion in the third direction is also neglected because of the small bandwidth in that direction.

The above dispersion relation has a different character at high and low frequencies. For $\omega^2 - \omega_e^2\Delta_0^2 \gg \alpha\omega_e^2$, $\omega(q)$ is nearly independent of q and the density of states is that of a 1D magnon spectrum given by

$$\rho(\omega) = \frac{1}{4\pi}\frac{\omega/\omega_e^2}{\left[(\omega/\omega_e)^2 - \Delta_0^2\right]^{1/2}}. \tag{6.77}$$

For $\omega^2 - \omega_e^2\Delta_0^2 \ll \alpha\omega_e^2$ the dispersion relation can be expanded in terms of $k_y y$ and

$$\omega(k) = \omega_e\{\Delta_0^2 + k_x^2 x^2 + \alpha k_y^2 y^2\}^{1/2}. \tag{6.78}$$

Consequently, the density of states in this limit is

$$\rho(\omega) = \frac{1}{4\pi}\frac{\omega}{\omega_e^2\alpha^{1/2}}. \tag{6.79}$$

An approximate form of the density of states, appropriate for both limits is given by

$$\rho(\omega) = \frac{1}{4\pi}\frac{\omega/\omega_e^2}{\left[(\omega^2/\omega_e^2) - \Delta_0^2 + \alpha\right]^{1/2}}. \tag{6.80}$$

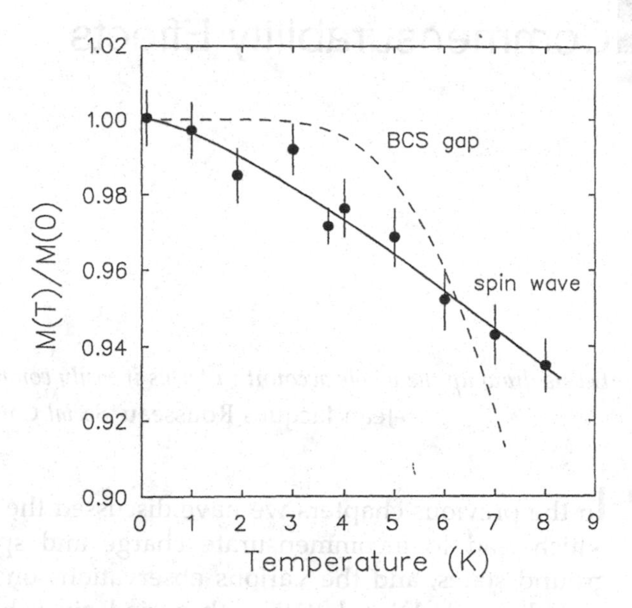

Figure 6.13. Temperature dependence of the magnetization $M(T)$ of $(TMTSF)_2PF_6$ as measured by muon spin rotation. The full line is Eq. (6.81) with parameters given in the text (Le et al., 1993).

The magnetization is given at temperature T by (Kittel, 1963)

$$M(T) = M_0 \int_{\omega_e \Delta}^{\omega_d} d\omega \frac{\rho(\omega)}{[(\omega^2/\omega_e^2) - \Delta_0^2]^{1/2}\left[\exp\left(\frac{\hbar\omega}{k_B T}\right) - 1\right]} \tag{6.81}$$

with the cutoff frequency defined as

$$\omega_d \int_{\omega_e \Delta_0}^{\omega_d} d\omega\, \rho(\omega) = \frac{1}{2}. \tag{6.82}$$

The magnetization calculated from Eq. (6.81) with the parameters established earlier is shown in Fig. 6.13, together with the magnetization as measured by muon spin rotation. At low temperatures, the temperature dependence is clearly determined by magnons and single particle excitations do not play an important role. Close to T_{SDW} both collective and single particle excitations may be important; but which of these determines the temperature dependence of the internal field remains to be seen.

7 Commensurability and Near Commensurability Effects

Let us draw up the whole account in terms if easily commensurable.
—Jean Jacques Rousseau *Social Contract*

In the previous chapters we have discussed the phase transitions which lead to incommensurate charge and spin density wave ground states, and the various observations on these states. The periodic underlying lattice with period given by the lattice constant a has been neglected, and its coupling to the periodic charge and spin density wave modulation (with period $\lambda_0 = \pi/k_F$) has been ignored. As a result, the ground state is translationally invariant; the ground state energy and the single particle gap are independent of the phase variable ϕ. Also the collective excitations related to the phase remain gapless (in the absence of Coulomb interaction as discussed in Chapter 6). This is all valid for relatively weak lattice potentials, such as those expected here.

It is anticipated that the effect of the underlying lattice periodicity can be ignored as long as the period of the charge or spin density wave modulation is rather different from the wavelength λ_0. However, if the two periods are commensurate, i.e.,

$$\lambda_0 = \frac{N}{M} a \tag{7.1}$$

where N and M are integers, or are nearly commensurate (with Eq. (7.1) close to being obeyed) the interaction between the two periodicities becomes important. This then has several important consequences. The ground state energy becomes a periodic function of the phase variable ϕ, and the phase excitations develop a gap in the $q \to 0$ limit. Because of the quasiperiodic dependence of the ground state on the phase, various nonlinear excitations occur; the most common of these, soliton solutions, have been thoroughly explored. The situation is somewhat different near commensurability, where the density wave can distort itself in order to develop commensurate regions, separated by discommensurations.

These effects can be described by representing the lattice by the interaction term

$$\mathscr{H} = V_0\{1 - \cos(M\phi)\} \tag{7.2}$$

where, for simplicity, we assume that in Eq. (7.1) $N = 1$. The potential V_0 can be derived from a microscopic theory using weak coupling arguments and it decreases with an increasing degree of commensurability M. Having established the magnitude of V_0, the term given above can be added to the Lagrangian, which has been used before; and the excitations of the system can be described within the framework of the Ginzburg-Landau theory. This approach will be taken here.

Whether such commensurability effects can be seen in the materials which have been discussed before remains to be seen. Structural studies conducted in several materials, as well as electron counting arguments; suggest that at least some of the compounds are commensurate or close to commensurability. In these cases $\lambda_0 \simeq 4a$. As will be discussed, for such higher order commensurability the potential V_0 is small, and consequently the energy associated with the periodic variation of the phase is also small and can be easily overshadowed by other effects such as interactions with impurity potentials. Nonlinear excitations have also been searched for by a variety of optical experiments; clear evidence for such states, however, has not been found.

7.1 Models of Commensurability Effects

As discussed in Chapters 3 and 4, the condensation energy was found to be independent of the phase variable ϕ. The reason for this is that by approximating the dispersion of the electron gas by the expression $\epsilon_k - \epsilon_F = \hbar v_F(k - k_F)$ the underlying lattice periodicity is ignored. From Eq. (3.27) the equation of motion, is

$$\begin{aligned} E_k \cdot a_{1,k} &= \epsilon_k a_{1,k} + \Delta e^{-i\phi} a_{2,k} \\ E_k \cdot a_{2,k} &= -\epsilon_k a_{2,k} + \Delta e^{i\phi} a_{1,k} \end{aligned} \tag{7.3}$$

where only states on the left and right hand side of the Fermi surface are included. The new dispersion relation is obtained from

$$\begin{vmatrix} E_k - \epsilon_k & -\Delta e^{-i\phi} \\ \Delta e^{i\phi} & E_k + \epsilon_k \end{vmatrix} = 0 \tag{7.4}$$

giving $E_k = \pm(\epsilon_k^2 + \Delta^2)^{1/2}$ as in Eq. (3.34). Here we have assumed that states near $+k_F$ couple to states near $-k_F$ (and vice versa). As usual, the periodic lattice also leads to a mixture of states which differ by $\pm 2k_F$, $\pm 4k_F, \cdots$; however, it is assumed that only the first term is important as the other states (except those included in Eq. (7.3)) are far from the Fermi energy. In the case of commensurate density wave, however, this is not the case, and the situation is illustrated in Fig. 7.1 for a half filled and for a one-third filled band. The former corresponds to $\lambda_0 = 2a$ the latter to $\lambda_0 = 3a$. In the case of the half filled band $a_{2,k-2k_F}$ is equivalent to $a_{1,k}$ and the equation of motion is (neglecting for the moment that a half filled band is a special case, and the order parameter is real)

$$\begin{aligned} E_k a_{1,k} &= \epsilon_k a_{1,k} + \Delta e^{-i\phi} a_{2,k} + \Delta e^{i\phi} a_{2,k} \\ E_k a_{2,k} &= -\epsilon_k a_{2,k} + \Delta e^{-i\phi} a_{1,k} + \Delta e^{i\phi} a_{1,k} \end{aligned} \tag{7.5}$$

giving

$$E_k = \pm\left(\epsilon_k^2 + 4\Delta^2 \cos^2\phi\right)^{1/2} = \pm\left\{\epsilon_k^2 + 2\Delta^2(1 + \cos(2\phi))\right\}. \tag{7.6}$$

The dispersion relation now depends on the phase ϕ, and so does the single particle gap which has a maximum when $\phi = 0$, here

$$\Delta = 2\epsilon_F e^{-1/2\lambda}. \tag{7.7}$$

For $\phi = \pi/2$, there is no single particle gap. For $\phi = 0$

$$E_k = \pm\left(\epsilon_k^2 + 4\Delta^2\right)^{1/2} \tag{7.8}$$

and the condensation energy (see Eq. (3.40)) is

$$E_{\text{el}} = n(\epsilon_F) \cdot 4\Delta^2 \, ln\left(\frac{2\epsilon_F}{\Delta}\right) = \frac{n(\epsilon_F) \cdot 2\Delta^2}{\lambda} \tag{7.9}$$

using Eq. (7.7). In general, the phase dependent electronic energy can be written as

$$E_{\text{el}} = \frac{n(\epsilon_F)\Delta^2}{\lambda} \cos 2\phi. \tag{7.10}$$

All this can be generalized to higher order commensurability. The potential as given by Eq. (7.1) connects electrons and holes with $\pm k_F$ in two different ways. There is a direct scattering from $-k_F$ to $+k_F$ as shown in Fig. 7.1. Alternatively, $M-1$ repeated scatterings each involving $-2k_F$ leads to the wavevector $2k_F(M-1)$. By symmetry however, this is equivalent to a scattering involving $+2k_F$ and therefore leads to the same final state. The process involves scattering into states which are far removed from ϵ_F, and this then leads to a reduction of the potential V_0. For a commensurability M one obtains

$$E_k = \pm\left(\epsilon_k^2 + 2\Delta^2 + \frac{2\Delta^M}{D^{M-2}} \cos M\phi\right)^{1/2} \tag{7.11}$$

where D is a cutoff energy, assumed to be on the order of the bandwidth or the Fermi energy (Lee, Rice, and Anderson, 1974). For $M=2$ we recover Eq. (7.8). Integrating over the occupied states leads to the commensurability energy

$$E_{\text{comm}} = -\frac{n(\epsilon_F)\Delta^2}{\lambda}\left(\frac{\Delta}{D}\right)^{M-2} \cos M\phi \tag{7.12}$$

which reduces to Eq. (7.10) in the $M=2$ case.

Having established the phase dependent condensation energy, this term can be added to the Lagrangian which describes the time and spatial dependence of the phase fluctuations, which were discussed in Chapter 6. The free energy, from Eq. (6.36), is given at $T=0$ by (with amplitude fluctuations neglected here)

$$F(0) = \frac{1}{4}n(\epsilon_F)\int dx\left[\hbar^2 v_F^2\left(\frac{d\phi}{dx}\right)^2 + \frac{m^*}{m}\left(\frac{d\phi}{dt}\right)^2 \right. \tag{7.13}$$

$$\left. + \frac{1}{M^2}\omega_F^2\frac{m^*}{m}\cos(M\phi)\right]$$

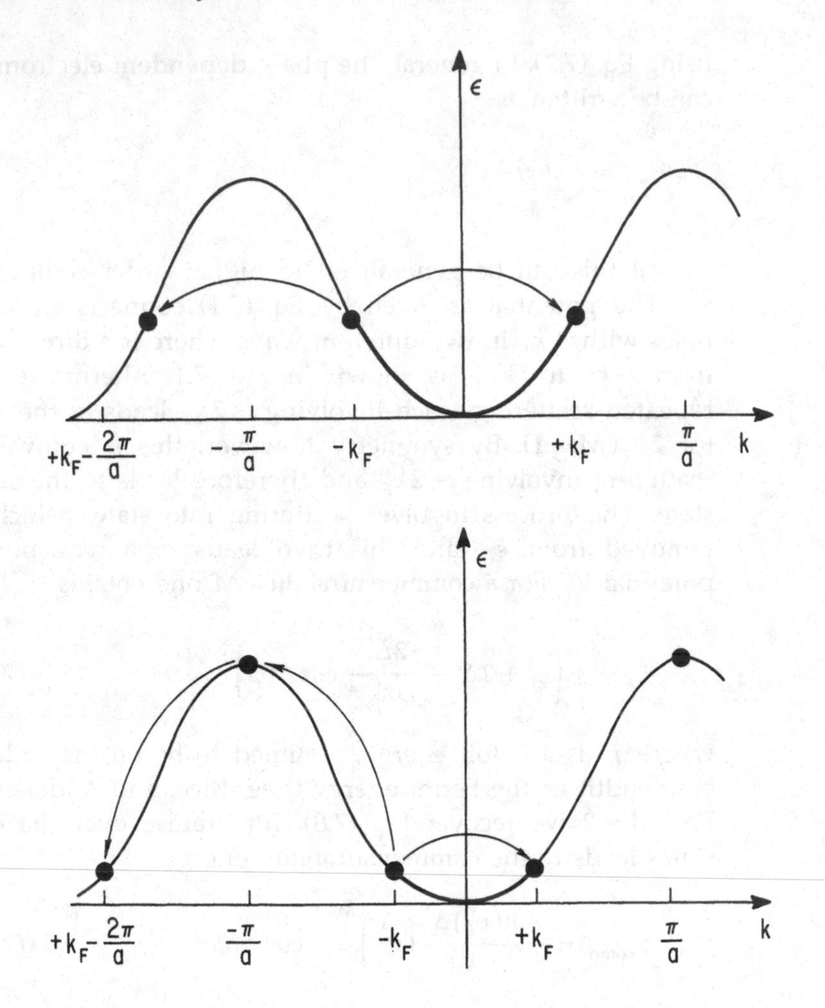

Figure 7.1. The dispersion relation for a half filled (top part) and one-third filled (bottom part) electron band. The scattering involving the wavevector $2k_F$ and the scatterings (one in the half filled band and two in the one-third filled band case) which lead to $\pm k_F - 2\pi/a$ are indicated on the figure by the arrows.

with the notation

$$\omega_F^2 = \frac{4M^2}{\lambda}\frac{m\Delta^2}{m^*}\left(\frac{\Delta}{D}\right)^{M-2}. \tag{7.14}$$

The Lagrange equation, Eq. (6.8a), gives the following equation of motion

$$\frac{d^2\phi}{dt^2} - \hbar^2 v_F^2 \frac{m}{m^*}\frac{d^2\phi}{dx^2} + \frac{\omega_F^2}{M}\sin(M\phi) = 0 \tag{7.15}$$

the well-known Sine-Gordon equation.

First we look at the small amplitude displacements of the undistorted density wave around the equilibrium position. Then the potential is

$$V(\phi) = \frac{1}{2}\frac{d^2V}{dx^2}\phi^2 = \frac{\omega_F^2}{2M}\phi^2. \tag{7.16}$$

Assuming that the time and spatially dependent phase is given by $\phi = \phi \exp[i(\omega t - qx)]$ we obtain the following dispersion relation

$$\omega_\phi(q) = \left(\frac{\omega_F^2}{M} + v_F^2\frac{m}{m^*}q^2\right)^{1/2} = \left(\frac{\omega_F^2}{M} + c_\phi^2 q^2\right)^{1/2} \tag{7.17}$$

where we have used the expression of the phase velocity given by Eq. (6.33). In contrast to Eq. (6.33) where the phase excitations are gapless, the periodic potential leads to a gap in the excitation spectrum, and the dispersion relation is shown in Fig. 7.2.

For arbitrary distortions of the phase, the solution $M\phi(x,t) = s(\xi)$ is obtained by performing the Galilean transformation $t \rightarrow$

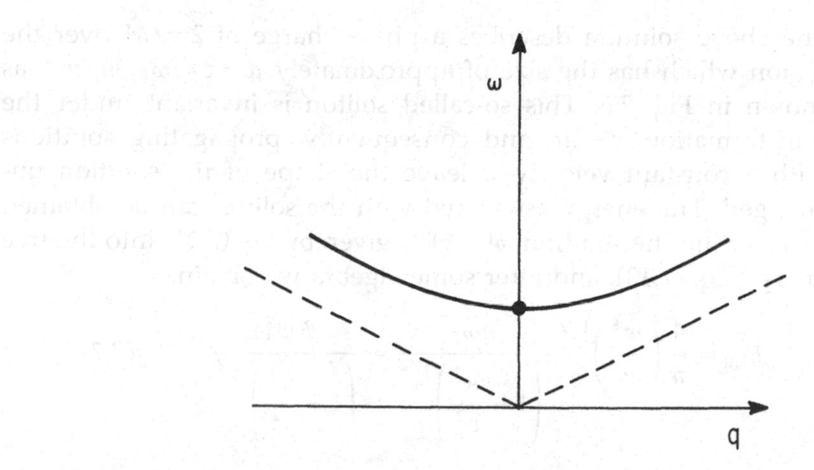

Figure 7.2. The phason dispersion relation for a commensurate (full line) and incommensurate (dashed line) density wave. The dispersion relations are given by Eqs. (7.17) and (6.33).

$t' = t$ and $x \to \xi = x - ut$. Then the integration with respect to ξ gives

$$\frac{1}{2}\left(c_\phi^2 - u^2\right)\left(\frac{ds}{d\xi}\right)^2 = -\frac{\omega_F^2}{M}\cos(s) + \text{const} \tag{7.18}$$

where we have the relations $\partial/\partial t = -u(\partial/\partial\xi)$ and $\partial/\partial x = d/d\xi$. Choosing the constant, $c = \omega_F^2/M$ we obtain

$$M\frac{ds}{d\xi} = \pm\frac{4\omega_F/M^{1/2}}{\left(c_\phi^2 - u^2\right)^{1/2}}\sin\left(\frac{s}{4}\right)\cos\left(\frac{s}{4}\right) \tag{7.19}$$

and the integration gives

$$\tan\left(\frac{Ms}{4}\right) = \exp\left\{\pm\frac{\omega_F/M^{1/2}}{\left(c_\phi^2 - u^2\right)^{1/2}}\right\}. \tag{7.20}$$

Therefore,

$$\phi^\pm(x,t) = \frac{s(\xi)}{M} \tag{7.21}$$

$$= \frac{4}{M}\tan^{-1}\left[\exp\left\{\pm\frac{\omega_F/M^{1/2}}{\left(c_\phi^2 - u^2\right)^{1/2}}(x - ut)\right\}\right].$$

The above solution describes a phase charge of $2\pi/M$ over the region which has the size of approximately $d = v_F^2/\omega_F, m/m^*$, as shown in Fig. 7.3. This so-called soliton is invariant under the transformation $x - ut$, and consequently, propagating solutions with a constant velocity u leave the shape of the solution unchanged. The energy associated with the soliton can be obtained by inserting the solution $\phi(x,t)$ as given by Eq. (7.21) into the free energy Eq. (7.13), and after some algebra we obtain

$$E_{\text{sol}} = \frac{4}{\pi}\left(\frac{m^*}{m}\right)^{1/2}\frac{\hbar\omega_F}{\left(1 - \dfrac{u^2}{c_\phi^2}\right)^{1/2}} = \frac{Mc_\phi^2}{\left(1 - \dfrac{u^2}{c_\phi^2}\right)^{1/2}}. \tag{7.22}$$

The creation of a soliton leads to a change of phase by $2\pi/M$ going (say) from the left hand to the right hand side of the soliton. As there is $2e$ charge associated with a density wave

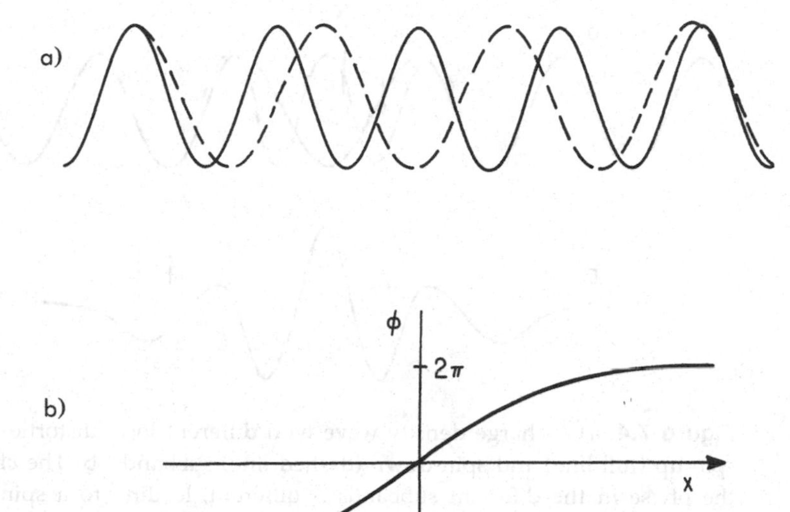

Figure 7.3. a) A density wave with a local distortion, as given by Eq. (7.21) (full line). The dashed line indicates the undistorted density wave. b) The spatial dependence of the phase ϕ.

period λ_0 (which corresponds to a phase $\pi/2$), a soliton excitation corresponds to a charge

$$e^* = \frac{2e}{M}. \tag{7.23}$$

Obviously, solitons with a change of phase by $2\pi n/M$, where n is an integer, can also be created, and the excitations correspond to an effective charge $2en/M$. In general, we can show that the charge associated with any local configuration leading to the change of phase is given by

$$e^* = \frac{\phi(\infty) - \phi(-\infty)}{\pi}. \tag{7.24}$$

Until now we have considered only one electron band, neglecting the spin degrees of freedom. Consequently, only charged excitations can be created by creating a local distortion of the phase $\phi(x,t)$. Let us now assume that the charge density wave is formed by two bands, one with spin up and one with spin down.

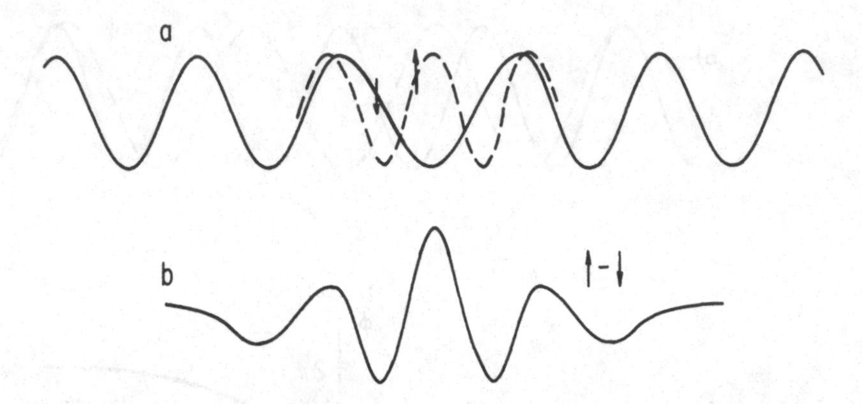

Figure 7.4. a) A charge density wave with different local distortions in the spin up (full line) and spin down (dashed line) subbands. b) The change of the phase in the different subbands is different, leading to a spin density modulation (↑ − ↓), as shown.

A soliton excitation $\phi(x,t)$ which is the same in both spin subbands returns our previous results and only charge excitations with charge $e^* = 2\pi/M$ occur. However, the form of the localized excitation can be different in the two subbands, generally leading to both charge and spin excitations. A simple example of such excitation (which may occur in response to an impurity potential) is shown in Fig. 7.4; where the curve shows the two soliton excitations of the two subbands, and the lower curve corresponds to the difference of the two charge excitations in the two subbands, i.e., a localized spin excitation.

The counting rule, Eq. (7.24), can be generalized by defining the following phases

$$\phi(x) = \frac{\phi\downarrow(x) + \phi\uparrow(x)}{2}$$
$$\theta(x) = \frac{\phi\uparrow(x) - \phi\downarrow(x)}{2}. \tag{7.25}$$

In terms of ϕ and θ,

$$e^* = \frac{\phi(\infty) - \phi(-\infty)}{\pi}$$
$$S_z = \frac{\theta(\infty) - \theta(-\infty)}{\pi}. \tag{7.26}$$

For the excitation, such as that shown in Fig. 7.4, $\phi(\infty) = \phi(-\infty)$ and therefore the change associated with the excitation is zero. However, the phase change is different in the two subbands, and $\theta(\infty) - \theta(-\infty) = 2\pi$. This leads to a spin $S = 1/2$ which is associated with the excitation. The above also suggests that for arbitrary phases fractional changes and fractional spins may be created, the notion of a fractional spin however is not universally accepted (see, for example, Horowitz (1989)).

The periodic lattice potential also leads to important, but somewhat different, effects when the density wave is not entirely, but only nearly, commensurate with the underlying lattice. For a commensurate and undistorted density wave the phase is constant. For a nearly commensurate density wave, in contrast, the local phase varies slowly with position and this variation is described as

$$\phi_l = \phi_0 + qx \tag{7.27}$$

where q accounts for the deviation from commensurability and is written as

$$q = \frac{2\pi}{\lambda_0}(Ma - \lambda_0). \tag{7.28}$$

The free energy is then written as

$$F(0) = \frac{1}{2}n(\epsilon_F)\int dx \left[v_F^2 \left(\frac{d\phi}{dx} - q \right)^2 + \frac{m^*}{m}\left(\frac{d\phi}{dt} \right)^2 + \frac{\omega_F^2}{M}\frac{m^*}{m}\cos(M\phi) \right]. \tag{7.29}$$

The solution is a periodic array of solitons, separated by regions where the density wave is commensurate with the underlying lattice. The situation is sketched in Fig. 7.5 where the linear variation of ϕ_ℓ corresponds to an incommensurate density wave. The distance between the solitons, ℓ_s, follows from the simple consideration of the phase which must advance at the same rate for the soliton lattice as for the incommensurate density wave. As each soliton advances in phase by $2\pi/M$, the number of solitons over a distance x is $n = qx/(2\pi/M)$, and the distance

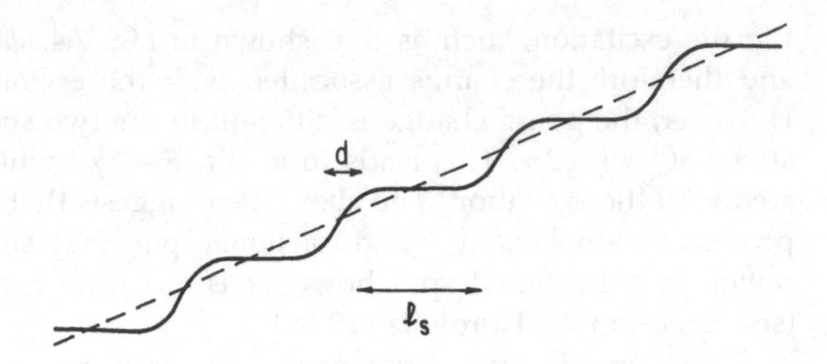

Figure 7.5. Variation of the local phase, ϕ_ℓ, for an incommensurate charge density wave (dotted line); and for a density wave which, due to interaction with the lattice, has commensurate regions, separated by discommensurations (full line). The distance between the solitons of with d is ℓ_s.

between them is

$$\ell_s = \frac{x}{n} = \frac{2\pi}{qM}. \tag{7.30}$$

Rather near to commensurability, $q \ll 1$, the distance between solitons is significantly larger than their width. In this limit $\phi(x,t)$ for each individual soliton is given by Eq. (7.22) and their interaction can be neglected.

Assuming the incommensurate density wave having zero energy (except of course the condensation energy), the energy of the soliton lattice is given by two terms. The first is the commensurability energy, which over the distance ℓ_s is $\ell_s E_{\text{comm}}$. The second term, given by E_s, is the one associated with the solitons. Consequently, the total energy (per soliton) is

$$\begin{aligned} E_{\text{tot}} &= \ell_s\, E + E_{\text{sol}} \qquad (7.31) \\ &= \frac{-2\pi}{Ma - \lambda_0} \frac{n(\epsilon_F)}{4M^2} \frac{m^*}{m} \hbar^2\omega_F^2 + \frac{4}{\pi M^2}\left(\frac{m^*}{m}\right)^{1/2} \hbar\omega_F \end{aligned}$$

which after some algebra and using the expression for ω_F becomes

$$E_{\text{tot}} = \frac{\hbar\omega_F}{M^2}\left(\frac{m}{m^*}\right)^{1/2}\left\{-\frac{\pi M}{Ma - \lambda_0}\frac{n(\epsilon_F)\Delta}{\lambda^{1/2}}\left(\frac{\Delta}{D}\right)^{\frac{M-2}{2}} + \frac{4}{\pi}\right\}. \tag{7.32}$$

For a small deviation for commensurability $Ma - \lambda_0$ is small, the first term dominates, and the soliton lattice is stable. With increasing $Ma - \lambda_0$, a transition to an incommensurate density wave occurs. The approximate condition for this is given by $\ell E_{comm} = E_s$. For $M = 4$ this leads to

$$4a - \lambda_0 = \frac{2\pi^2}{\lambda^{1/2}} \frac{E_{cond}}{D} \tag{7.33}$$

where $E_{cond} = \frac{1}{2} n(\epsilon_F)\Delta^2$ is the condensation energy.

This crude argument neglects the (repulsive) interaction between solitons. This will become increasingly important as their separation becomes comparable to their spatial extension, which occurs near the soliton lattice-incommensurate DW transition. A sharp phase transition to an ordered state is not expected to occur in such a strictly one-dimensional system and arguments discussed in Chapter 5 also apply here. Interchain interactions, mediated either by Coulomb effects or by interchain tunneling alter this picture dramatically. Due to such interchain interactions we expect soliton arrays perpendicular to the chain direction, with the formation of such ordered structures leading to a phase transition from a 3D incommensurate density wave state to a 3D soliton lattice (or other discommensuration lattice) state.

In discussing the effects caused by the underlying periodic lattice potential we have assumed that the amplitude of the density wave is constant; consequently only variations of the phase variable have been considered. The nonlinear excitations are then called phase-solitons. Nonlinear excitations due to local changes of amplitude also occur, and such so-called amplitude solitons have also been widely discussed in the literature (see, for example, Bak and Pokrovsky, 1981 and references cited therein).

7.2 Experiments: Search for Commensurability Effects and Solitons

In several materials which have a charge or spin density wave ground state, the period of the density wave is commensurate or nearly commensurate with the underlying lattice. The wavevector is usually temperature dependent, and a typical dependence is shown for $K_{0.3}MoO_3$ in Fig 7.6. In this case the value $1 - q = a'$ refers to the wavevector which appears in the CDW modulation

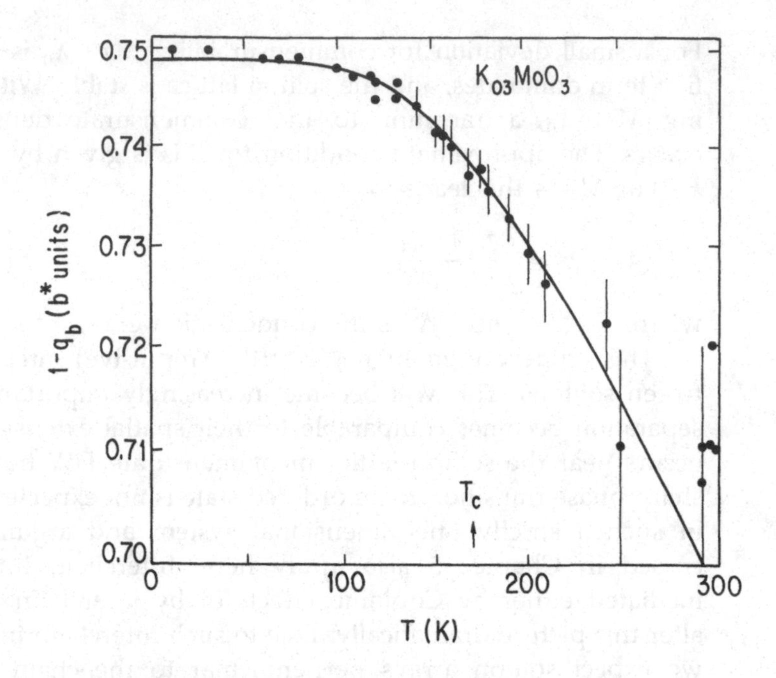

Figure 7.6. Temperature dependence of the charge density wavevector in $K_{0.3}MoO_3$. The notation used gives a CDW period $\lambda_0 = 4a$ (i.e., a four times commensurability) for $1 - q_b = 0.75$. After J. P Pouget and R. Comes (1989).

which has the form of $\cos(q'x + \phi)$, and thus at low temperatures the period is close to four times the lattice constant. The situation is similar for TaS_3 and also in $NbSe_3$. In $(TMTSF)_2PF_6$ the spin density wave is commensurate along the conducting chains, with $\lambda_0 = 4a$, and most probably this is the case for the other members of this group of materials.

Because of this, commensurability and near commensurability effects have been searched for in these materials; clear evidence for these effects, such as a gap in the phason excitation spectrum or nonlinear excitations at energies below the single particle gap, have not been found.

The most likely reason for this is that at or near commensurability the pinning potential is small. For a typical material, such as $K_{0.3}MoO_3$, for which $\Delta = 700$ K, $m^*/m \simeq 400$ and $\lambda \simeq 0.5$ (see Chapter 3) the pinning energy from Eq. (7.14)

$$\hbar\omega_F \simeq 2 \times 10^{-2}\Delta \simeq 10^{-3}\,\text{eV},$$

is fairly small. As will be discussed in Chapter 8, the interaction of density waves with impurities also leads to a pinning of the collective mode, and the relevant energy is larger than that given above. Consequently, it is likely that impurity rather than commensurability effects are responsible for the finite frequencies observed for $q \rightarrow 0$ in various materials. This has been directly confirmed by experiments conducted on alloys of various materials. The soliton energy for the same set of parameters as used before is

$$E_s = \frac{4}{\pi}\left(\frac{m^*}{m}\right)^{1/2} \hbar\omega_F \simeq 3 \times 10^{-2}\,\text{eV}$$

which is somewhat smaller than the single particle gap. The spatial extension of the soliton

$$d = \frac{(m/m^*)^{1/2} v_F}{\omega_F} \simeq 30\,\text{Å}$$

for $v_F \simeq 10^7$ cm/sec. These excitations, which involve the change of the phase of the collective mode, are expected to be optically active, and they have been searched for by a variety of optical probes without success. It has also been proposed that solitons contribute to the frequency and electric field dependent conductivity (Horowitz and Trullinger, 1984), however experimental evidence for such contributions has not yet been found.

8 The Interaction Between Density Waves and Impurities

...it is difficult to reconcile the interests of the conflicting sides.
—Mikhail Gorbachev *Perestroika*

Impurities which are distributed at random in the crystal have a profound influence on both the static and dynamic properties of density waves. Isoelectronic impurities (if they do not lead to a local lattice-distortion) do not couple to density waves while the electrostatic impurity potential $V(\vec{r}-\vec{R}_i)$ of the nonisoelectronic impurities leads to changes of both the amplitude and phase of the condensate near the impurity site $\vec{R}_i$. These changes have been discussed in terms of microscopic models, and the resulting spatial dependence of the charge and spin distribution has been evaluated. For small impurity potentials the amplitude of the density waves is only weakly modified, while for strong impurities (which can be represented by electrostatic potentials larger than the single particle gap) the amplitude collapses near $\vec{R}_i$ leading to bound states. The spatial extension of these states is on the order of the coherence length ξ, at zero temperature. For distances $|\vec{r}-\vec{R}_i| > \xi$ the influence of impurities can be described in terms of an interaction where the impurity potential couples directly to the phase ϕ of the condensate and the amplitude remains unaffected. The problem of many impurities distributed at random can be then treated within the framework of the Ginzburg-Landau description (Fukuyama and Lee, 1978; Lee and Rice, 1979).

Impurity potentials which couple directly to the phase ϕ of the condensate, destroy long-range order (Imry and Ma, 1975;

Sham and Patton, 1976; Efėtov and Larkin, 1977) and lead to a finite phase-phase correlation length L_0. The length scale depends on the strength of the impurity potentials, on the impurity concentration, and on the properties of the density waves. The absence of long range order, and the appearance of many metastable states has been confirmed by a variety of structural, thermodynamic and electric measurements.

8.1 Theories of Density Wave-Impurity Interaction

The usual treatment of the problem deals with impurity effects at low temperatures where temperature induced fluctuations as well as quantum fluctuations of the collective mode are neglected. The Hamiltonian which describes the interaction is given within the framework of the Ginsburg-Landau theory by (see Eq. (6.36))

$$\mathcal{H} = \frac{\hbar^2 v_F^2 n(\epsilon_F)}{2} \int d\vec{r} \left(\vec{\nabla}\phi\right)^2 + \frac{\pi}{2} V_{\text{imp}}(\phi) \tag{8.1}$$

where anisotropy effects due to the anisotropic band structure are neglected. These can be included by renormalizing the length scales in the different orientations, as was done in Chapter 6. The first term is the normal elastic energy (see Eq. (6.51)) and the second describes the interaction between the collective mode and impurities. The approach is appropriate for long wavelength deformations; short wavelength modifications of the phase on the length scale $L < \xi_0 = \hbar v_F / \pi\Delta$ cannot be discussed with the framework of this description.

The mechanism which leads to an interaction between charge density waves and impurities is shown in Fig. 8.1. The impurity potential $V(\vec{x}) = V_0\ \delta(\vec{r} - \vec{R}_i)$ at $\vec{R}_j$ induces a charge density oscillation with period $\lambda_0 = \pi / k_F$ (the well known Friedel oscillation) which is phase-matched to the density wave at position $\vec{R}_i$. For small impurity potentials this interaction can simply be written as

$$V_{\text{imp}}(\phi) = \int d\vec{r}\, V\left(\vec{r} - \vec{R}_i\right) \rho_1 \cos\left(\vec{Q}\vec{R} + \phi\left(\vec{R}_i\right)\right) \tag{8.2}$$

where ρ_1 is the amplitude of the charge density wave modulation. Obviously this is appropriate only if the impurity potential does not lead to modification of the amplitude of the collective

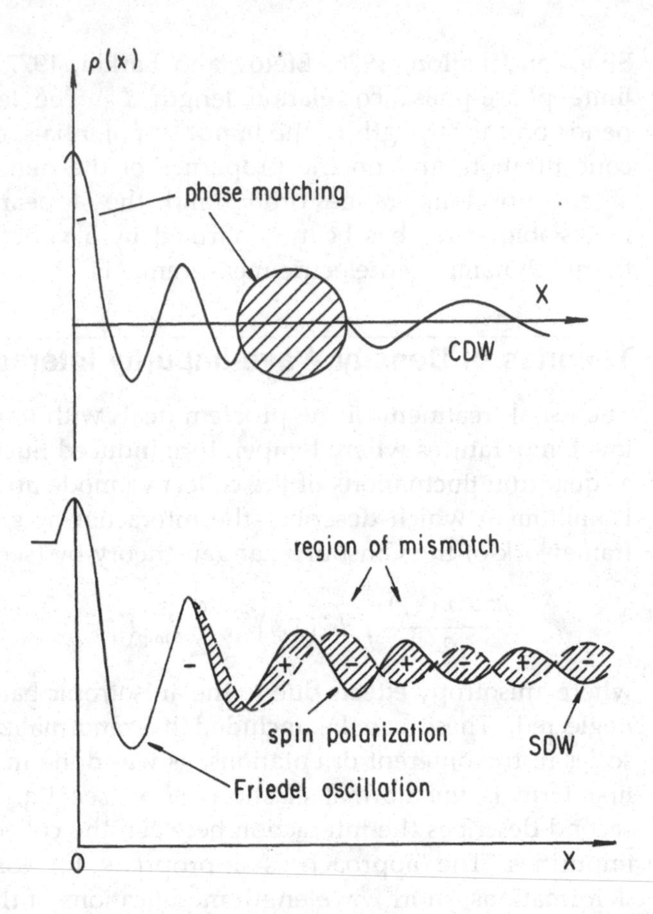

Figure 8.1. The local density distribution around a nonmagnetic impurity for a CDW and a SDW ground state (Tüttő and Zawadowski, 1985).

mode. This only occurs if V_0 is small. If V_0 is comparable to or larger than the single particle gap, bound impurity states may occur. These have been discussed in detail by Tüttő and Zawadowski (1985).

In the case of spin density waves, the situation is different. In the first order, charged impurities do not interact with the ground state. However, in second order the interaction between the impurity and the two subbands (both with a modulation of the charge density) is different (Fig. 8.1), leading to an interaction energy given by

$$V_{\text{imp}}(\phi) = \int d\vec{r}\, V\left(\vec{r} - \vec{R}_i\right) \rho_1 \cos\left(\vec{Q}_F \cdot \vec{R}_i + 2\phi(R_i)\right). \quad (8.3)$$

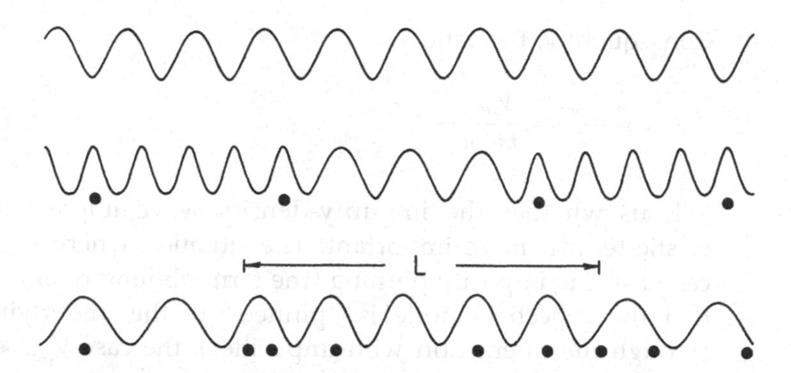

Figure 8.2. Strong and weak impurity pinning in one dimension. The phase is fully adjusted at each impurity for strong pinning; for weak pinning such phase adjustment occurs over a characteristic distance L_0.

Here ρ_1 represents the charge density wave modulations in the two spin subbands; as discussed in Chapter 4 these are displaced by $\phi = \pi$ with respect to each other.

In Eq. (8.1) the first term favors a uniform phase, $\phi = \text{const}$, with distortions leading to an increase of the elastic energy due to the internal deformation of the collective mode. The second term however, favors local distortions of the phase so that ϕ is adjusted to the impurity potential, thus decreasing the electrostatic interaction energy as given by Eqs. (8.2) and (8.3).

The various situations which may occur can easily be seen for a one-dimensional density wave interacting with randomly positioned impurities, such as that shown in Fig. 8.2. For an attractive potential at site $\vec{R}_i$, the phase $\phi(\vec{R}_i) = 0$ leads to the maximum charge density at the impurity site and if this happens, the potential energy gain with respect to the potential experienced by the undisturbed density wave with random phase $\phi(\vec{R}_i)$ at the impurity positions is approximately given by

$$V_{\text{pot}} = V_0 \rho_1 n_i. \tag{8.4}$$

This adjustment however, would lead to an increased elastic energy, and from Eq. (8.1) this is approximately given by

$$V_{\text{el}} \approx \hbar v_F \left(\frac{2\pi}{<\ell>} \right)^2 \simeq \hbar v_F n_i^2, \tag{8.5}$$

where $<\ell>$ is the average distance between the impurities.

Consequently, the ratio

$$\epsilon = \frac{V_{\text{pot}}}{V_{\text{el}}} = \frac{V_0 \rho_1}{\hbar v_F n_i} \tag{8.6}$$

tells us whether the impurity-density wave interaction or the elastic term is more important. The situation where $V_{\text{pot}} \gg V_{\text{el}}$ is called strong impurity pinning (the term pinning refers to the fact that the collective mode is "pinned" to the underlying lattice through the interaction with impurities); the case $V_{\text{pot}} \ll V_{\text{el}}$ represents weak impurity pinning.

For both cases long range phase coherence is lost. For large distances we can write

$$\langle \phi(\vec{r}), \phi(0) \rangle \simeq e^{-\vec{r}/L_0} \tag{8.7}$$

with L_0 the so called phase-phase correlation length.

For $\epsilon \gg 1$, the potential energy term dominates and the total energy of the coupled collective mode and impurity is given by

$$E_{\text{tot}} = V_0 \rho_1 n_i. \tag{8.8}$$

As the phase is completely adjusted to the impurity positions at every impurity site, the average phase coherence length $L_0 \simeq \frac{1}{r_i}$ is the average distance between impurities.

The case for $\epsilon \lesssim 1$ is more interesting and in this limit scaling arguments can be used to evaluate the total energy and the phase-phase coherence length. In contrast to what happens for strong impurity potentials, expect the phase to adjust to many randomly positioned impurities with a length scale over which the phase varies by one wavelength significantly exceeding the average distance between impurities. The situation is shown in Fig. 8.3. Assuming that the length scale over which such adjustment occurs is L_0, in d dimensions the volume is L_0^d, and statistical arguments lead to an overall decrease of the potential energy

$$E_{\text{pot}} = -V_0 \rho_1 \left(L_0^d n_i \right)^{1/2} L_0^{-d} = -V_0 \rho_1 n_i^{1/2} L_0^{-d/2}. \tag{8.9}$$

per volume.

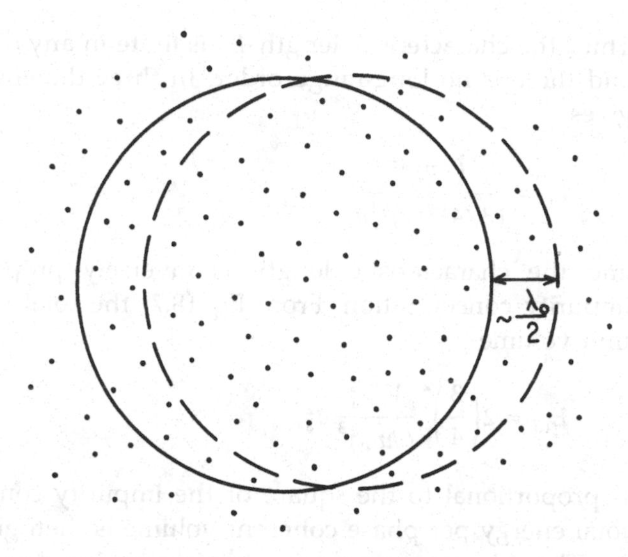

Figure 8.3. The weak impurity pinning in two dimensions. The full circle represents a region of undistorted density wave, the broken circle represents the density wave which is adjusted to optimize the interaction with the impurity potentials.

The change of the elastic energy, associated with such adjustment, comes from the gradient term in Eq. (8.1), and is given by

$$E_{el} \simeq \frac{\hbar v_F}{L_0^2}. \tag{8.10}$$

The total energy corresponding to the volume L_0^d is

$$E_{tot} = E_{pot} + E_{el} = -V_0 \rho_1 n_i^{1/2} L_0^{-d/2} + \frac{\hbar v_F}{L_0^2} \tag{8.11}$$

and by minimizing with respect to L_0 we obtain

$$-V_0 \rho_1 n_i^{1/2} \left(\frac{d}{2}\right) L_0^{-(d/2)-1} + 2\hbar v_F L_0^{-3} = 0 \tag{8.12}$$

or

$$L_0 = \left[\frac{dV_0 \rho_1 n_i^{1/2}}{4\hbar v_F}\right]^{+\frac{2}{d-4}}. \tag{8.13}$$

Thus, the characteristic length L_0 is finite in any dimension $d < 4$, and there is no long-range order. In three dimensions, Eq. (8.13) gives

$$L_0 = \frac{(\hbar v_F)^2}{(3/4)^2 V_0^2 \rho_1^2 n_i} \tag{8.14}$$

and the characteristic length is inversely proportional to the impurity concentration. From Eq. (8.7) the total energy gain per unit volume

$$E_{\text{tot}} = 2\left(\frac{3}{4}\right)^2 \frac{V_0^4 \rho_1^4}{(\hbar v_F)^3} n_i^2, \tag{8.15}$$

is proportional to the square of the impurity concentration. The total energy per phase coherent volume is then given by E_{tot}.

The above arguments can easily be generalized to an anisotropic band-structure, which we describe in terms of the anisotropic coherence length ξ, which is proportional to the Fermi velocity, see Eq. (6.17). Then the elastic term of the free energy reads

$$F_{\text{el}} = \frac{n(\epsilon_F)\pi^2}{16} \int d\vec{r} \left\{ \xi_x^2 \left(\frac{\partial \phi}{\partial x}\right)^2 + \xi_y^2 \left(\frac{\partial \phi}{\partial y}\right)^2 + \xi_z^2 \left(\frac{\partial \phi}{\partial z}\right)^2 \right\}. \tag{8.16}$$

With $dy' = (\xi_y/\xi_x)\, dx$ and $dz' = (\xi_z/\xi_x)\, dz$ Eq. (8.16) becomes

$$F_{\text{el}} = \frac{n(\epsilon_F)\pi^6}{16} \frac{\xi_y \xi_z}{\xi_x^2} \int d\vec{r}\, \xi_x^2 (\vec{\nabla}'\phi)^2 \tag{8.17}$$

with $dr' = dx'\ dy'\ dz'$. The volume which is evaluated becomes $V = L^d(\xi_y \xi_z/\xi_x^2)$ instead of L_0^d. With this rescaling, Eq. (8.10) becomes

$$E_d = \hbar v_F\, L_0^{-2} \left(\frac{\xi_y \xi_z}{\xi_x}\right)^2 \tag{8.18}$$

and the potential energy, from (8.9) becomes

$$E_{\text{pot}} = -V_0 \rho_1 \left(\frac{\xi_y \xi_z}{\xi_x^2}\right)^{1/2} n_i^{1/2} L_0^{-d/2}. \tag{8.19}$$

Minimization of the total energy then gives a characteristic length

$$L_0 = \frac{(\hbar v_F)^2}{(3/4)^2 V_0^2 \rho_1^2} \frac{1}{n_i} \frac{\xi_x^2}{\xi_y \xi_z} \tag{8.20}$$

and a total pinning energy

$$E_{\text{tot}} = 2\left(\frac{3}{4}\right) \frac{V_0^4 \rho_1^4}{(\hbar v_F)^3} n_i^2 \left(\frac{\xi_x^2}{\xi_y \xi_z}\right). \tag{8.21}$$

The characteristic phase-phase correlation length along the chain direction is L_0, and in the perpendicular directions is given by

$$L_y = \left(\frac{\xi_y}{\xi_x}\right) L_0$$

$$L_z = \left(\frac{\xi_z}{\xi_x}\right) L_0. \tag{8.22}$$

Some numerical estimates of the characteristic lengths and total energies per domain are in order here. The impurity potential cannot greatly exceed the single particle gap, and therefore $V_0 \simeq 5 \times 10^{-2}$ eV for a typical charge density wave gap. The amplitude of the CDW modulation $\rho_1 \simeq 10^{-2}$ for $v_F = 10^7$ cm/sec, leads to $\epsilon \sim 10^{-1}$ (see Eq. (8.6)) for $c = 1000$ ppm. For this impurity concentration the weak impurity pinning limit applies. With the above parameters, the phase-phase correlation length is $L_0 = 3 \times 10^{-3}$ cm along the chain direction, and significantly smaller perpendicular to the chains. The total energy per phase coherent volume from Eq. (8.21) is, for an anisotropy $\xi_y/\xi_x \sim 10^{-1}$ and $\xi_z/\xi_x \sim 10^{-2}$, on the order of 0.1 eV, significantly exceeding the thermal energy $k_B T$ well below the transition. Because of the smaller gap values, L_0 is larger and E_{tot} is smaller for materials with a spin density wave ground state.

The scaling arguments which lead to a finite correlation length for $d < 4$ are based on the assumption that phase excitations are gapless and can be described by the first term in Eq. (8.1). As discussed in Chapter 6, Coulomb interactions have a profound influence on the longitudinal fluctuations, while leaving the transverse modes unchanged. The consequence of this is that the collective mode can adjust to impurity potentials only through deformations which involve transverse distortions. Arguments similar to those advanced above lead to a critical dimension $d = 3$. Consequently, in the presence of Coulomb interactions, impurities do not destroy the long range order in the 3D ordered density wave state. The situation is somewhat more complicated at finite temperatures where screening due to normal electrons is important. This leads to a characteristic Thomas-Fermi wavevector $q_0 = 4\pi e^2 n_{qp}/T$, with n_{qp} a thermally excited quasiparticle density. This then defines a length scale $L_{qp} = q_0^{-1}$, and phase fluctuations remain gapless for length scales which exceed L_{qp}. Consequently, the scaling arguments which lead to Eqs. (8.19) and (8.15) are valid as long as $L_0 > L_{qp}$, but are expected to be significantly modified where L_0 as given by these equations is smaller than the Thomas-Fermi screening length.

8.2 Experimental Evidence for Finite Correlation Lengths

Although the theory outlined earlier has often been used to describe various experimental findings, many complications may arise. These complications are most probably responsible for the diversity of observations. The collective mode can be described as an anisotropic elastic medium only in the limit where the coherence length ξ significantly exceeds the lattice constants both parallel and perpendicular to the chains. While this may occur for some materials, such as $NbSe_3$, in the majority of cases $\xi < d$ at least for one direction. If this happens, the problem will likely resemble a weakly coupled system of layers, with E_{tot} and L_0 given for each by the two-dimensional variant of the Ginzburg-Landau theory. This may lead to concentration dependences different from those given before.

The phase-phase correlation length L_0 is usually long, a crude estimate gives L_0 of a few microns to millimeters in relatively pure specimens. L_0 is somewhat smaller in directions perpendicular to the chains due to anisotropy effects. These length scales are comparable to the dimensions of the specimens, and in the limit of $L_0 > L_{specimen}$ the low wavelength deformations of the collective mode are removed. This again modifies the concentration dependence of the characteristic energies and length scales and also suggests that some experimental findings may depend on the dimensions of the specimens investigated. Additional effects may be caused by surface effects, which may also contribute to E_{tot} in certain cases.

The absence of long-range order, as suggested by the theories discussed above, has several important effects on the static and dynamic properties of density waves. The finite phase-phase coherence length leads to the broadening of the X-ray diffraction lines which occur as a consequence of the charge density wave. Similar phenomena are implied for spin-density waves. In this case the structure factor for magnetic scattering would be modified. High resolution X-ray studies on materials in which impurities have been introduced by alloying or irradiation have been conducted in a few cases, and the absence of long range order is well established (Forro et al., 1983; Girault et al., 1988; Tamegai et al., 1987; DeLand et al., 1991; Sweetland et al., 1990).

The finite coherence length leads to broadening of the X-ray diffraction peaks, and the satellite peak profiles corresponding to the CDW wavevector $\vec{Q}$ can be written as

$$I^{\pm}(\vec{k}) \simeq \frac{\Gamma}{\left(\vec{k}-\vec{Q}\right)^2 + \Gamma^2}. \tag{8.23}$$

The half-width $\Gamma = 1/L_0$, and therefore gives the phase-phase correlation length L_0 directly. The X-ray intensity profile can be examined in different crystallographic directions, allowing for the measurement of the correlation lengths in different orientations. Such studies have been performed on alloys of $NbSe_3$ (Sweetland et al., 1990), and of $K_{0.3}MoO_3$ (DeLand et al., 1991). In both cases the coherence length was found to be anisotropic, with the anisotropy as expected for the anisotropic band structure. The concentration dependence of the width (referred to as full width at half maximum, FWHM) is displayed in Fig 8.4 for three orientations. The measured width is strongly anisotropic due to the anisotropic correlation length. The concentration dependence is found to be

$$L_0 \sim \frac{1}{n_i^{1/2}}, \tag{8.24}$$

in clear contrast to the consequences of weak impurity pinning which predicts that L_0 is inversely proportional to the impurity

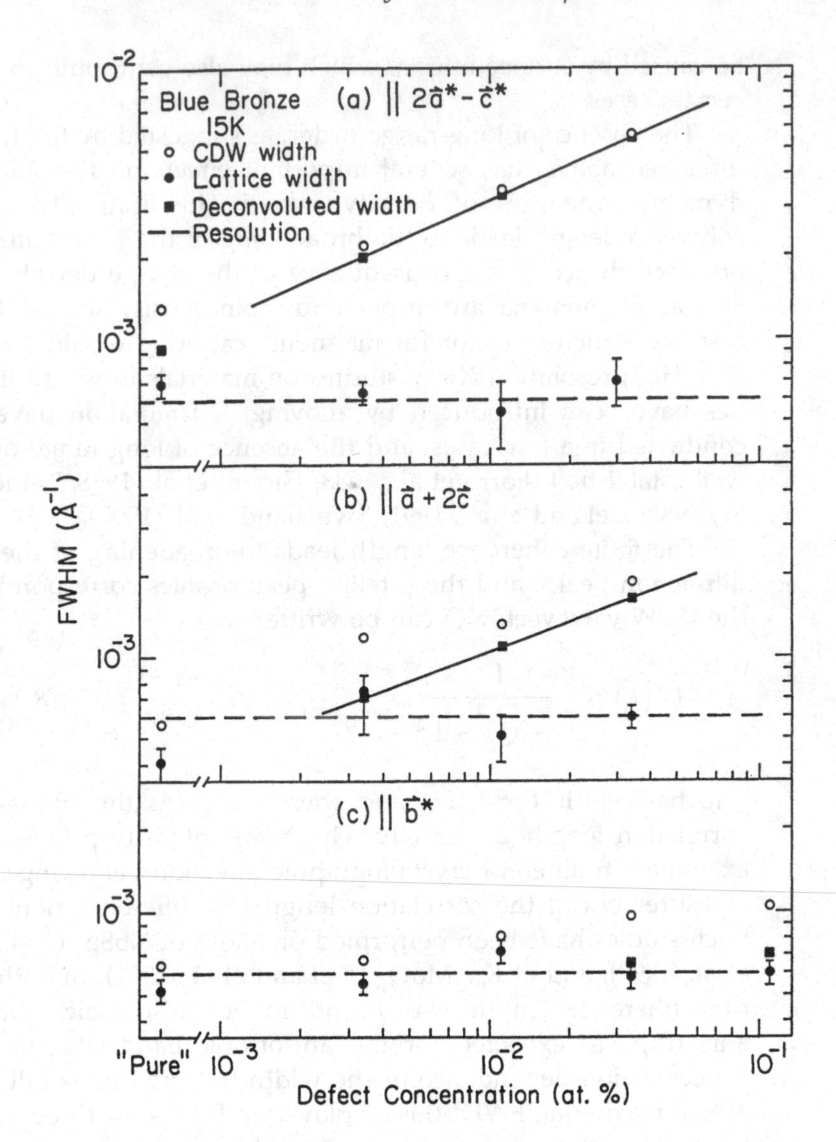

Figure 8.4. CDW satellite peak widths, widths of nearby lattice peaks, and deconvoluted widths versus defect concentration in $K_{0.3}MoO_3$. Widths are given as full widths at half maximum (FWHM). The CDW and lattice peaks are: (a) (5, 0.75, −2.5) and (4, 0, −2), (b) (2, −0.25, 1.5) and (2, 0, 1), and (c) (1, 2.75, −0.5) and (0, 2, 0). The resolution is shown as a dashed line (after DeLand, Mozurkewich, and Chapman, 1991).

concentration see (Eq. (8.20)). This disagreement between theory and experiment is not understood at present.

The disorder and the finite coherence length also lead to a distribution of energy levels for the various pinned density wave configurations, with thermally induced transitions between those configurations. This leads to a low temperature specific heat which is similar to that observed in various glasses and disordered systems (Biljakovich et al., 1986; Brown et al., 1988). For transitions between different configurations differing by an energy ϵ, the specific heat is given by the expression

$$C_V = k_B\left(\frac{\epsilon}{k_BT}\right)\operatorname{sech}^2\left(\frac{\epsilon}{k_BT}\right) \tag{8.25}$$

which, for a broad distribution of energy levels up to a cutoff energy ϵ_0, gives, for $k_BT \ll \epsilon_0$,

$$C_V = N_0 k_B T \tag{8.26}$$

where N_0 is the density of energy levels per unit energy. The low temperature specific heat, measured in various materials with a CDW ground state, is indeed large, and can approximately be described by Eq. (8.26). The experimental results are displayed in Fig. 8.5 together with a prototype glass, vitreous silicon. While these experiments clearly point to the disordered, glassy behavior of the condensate, which is pinned by randomly positioned impurities, no attempt has been made to relate the observed anomalies to the fundamental parameters of the problem, such as the phase-phase correlation length or the average pinning energy.

A more significant consequence of impurity-density wave interactions is the pinning of the collective modes to the underlying lattice. Due to this interaction charge and spin density waves cannot move freely in the lattice; but assume a well defined equilibrium position which is determined by the spatial distribution of impurities. A displacement out of this equilibrium position leads to a change in the collective mode-impurity interaction energy. For a rigid displacement of the collective mode, this is given by

$$E = -k(x - x_0)^2 \tag{8.27}$$

where x_0 refers to the equilibrium position and k is a restoring

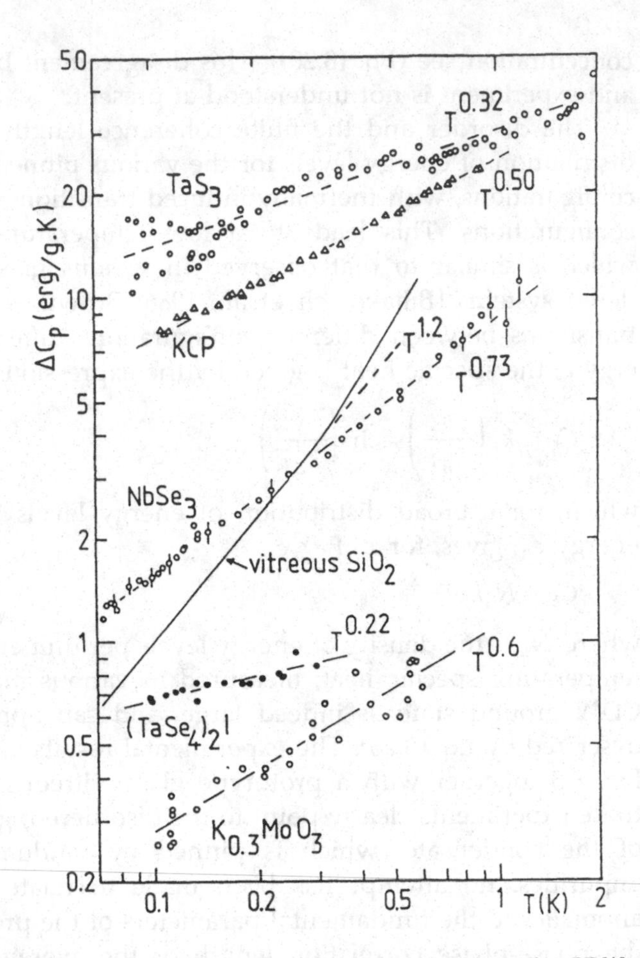

Figure 8.5. Residual specific heat of various materials in the CDW state. The phonon and nuclear hyperfine contributions to the specific heat have been subtracted from the data (after Biljakovich et al., 1989).

force. The latter can be estimated using the following crude argument. A rigid displacement of the (distorted) mode by roughly half the wavelength, $\lambda_0/2$, leads to a change in the impurity-collective mode interaction, and this change is on the order of the total pinning energy. Consequently,

$$k\left(\frac{\lambda_0}{2}\right)^2 \simeq E_{\mathrm{tot}} \tag{8.28}$$

with E_{tot} given by Eq. (8.8) or (8.15) for strong and weak impurity pinning. The restoring force therefore is given by

$$k \simeq \frac{4E_{tot}}{\lambda_0}. \tag{8.29}$$

Upon the influence of an applied *ac* field, the collective mode does not move freely. Due to the finite restoring force the oscillator frequency of the collective mode shifts from zero to a finite frequency. This frequency is well below the gap frequency $\omega_q = \Delta/\hbar$ for small impurity concentrations. Also, for small k a small *dc* electric field can induce a translational motion of the condensate, leading to sliding charge or spin density waves. These will be discussed in Chapters 10 and 11.

9 The Electrodynamics of Density Waves

Darkness with eyes open.
Light with eyes open.
Light with eyes shut.
The dial of life.

—Milan Kundera *Immortality*

Both charge and spin density wave condensates couple to electromagnetic fields and the fluctuations of the phase ϕ of the ground states lead to fluctuations of the electric current. In the absence of pinning due to impurities and lattice defects, the translational motion of the condensates will lead to a conductivity at zero frequency which, in the absence of damping, will also result in an infinite conductivity. This possibility was first raised by Fröhlich (1954). In addition to such effects related to the dynamics of the condensate, carrier excitations across the single particle gap lead to electromagnetic absorption, with an onset frequency of $\omega_g = 2\Delta/\hbar$ for the absorption process. These features are similar to those which are observed in superconductors; there are, however, several important differences which lead to electrodynamic properties fundamentally different from those of the superconducting ground state.

To the first order, the interaction of the collective modes and impurities can be represented as a restoring force k, leading to a collective mode contribution to the conductivity at finite frequency, which is approximately given by $\omega_0 \approx (k/m^*)^{1/2}$. The interaction of the collective modes with lattice vibrations and imperfections leads also to damping and therefore to a finite spectral width of the collective mode resonances. Furthermore, the dynamics of the internal deformations of the collective modes will lead to long time or low frequency relaxational modes, with a frequency dependent response similar to those observed in glasses and strongly disordered solids.

These aspects are well understood for charge density waves where extensive optical experiments clearly identified the single particle excitations and the contribution of collective mode resonances. The state of affairs is less clear for spin density waves, where major disagreement between theory and experiment remains.

9.1 The Electrodynamics of Density Waves

As in a superconductor, the phase $\phi(x,t)$ plays an important role in the dynamics of the collective modes. The $q=0$ phase mode corresponds to the translational motion of the condensed electrons, with the ions oscillating about their equilibrium positions. A rigid displacement of the density wave leads to an electric current, and the current density per chain $j_{\mathrm{DW}} = -n_{\mathrm{DW}} e v_d = -n_{\mathrm{DW}} e(dx/dt)$. With $\phi = 2k_F x$ and two electrons per density wave period $\lambda_0 = \pi/k_F$,

$$j_{\mathrm{DW}} = -\frac{e}{\pi}\frac{d\phi}{dt}. \tag{9.1}$$

A compression of the wave leads to a change of the electronic density, and therefore

$$n_{\mathrm{DW}} = \frac{e}{\pi}\frac{d\phi}{dx} \tag{9.2}$$

at zero temperature. The cross derivatives of the above equations lead to the continuity equation

$$\frac{dj_{\mathrm{DW}}}{dx} + \frac{dn_{\mathrm{DW}}}{dt} = 0. \tag{9.3}$$

The relation that connects the electric current and condensate density to the time and spatial derivative of the phase $\phi(x,t)$ of the condensate is different from the relations that are appropriate for a superconductor,

$$p_s = i\hbar \frac{d\phi}{dx} \text{ and } \mu = i\hbar \frac{d\phi}{dt} \tag{9.4}$$

where p_s and μ are the momentum and chemical potential, respectively.

The applied electric potential eE couples to the gradient of the phase, and the potential energy density is given by

$$H_E = \frac{eEx}{\pi} \frac{d\phi}{dx}. \tag{9.5}$$

Here (and also later in this chapter and in Chapter 10) the electric field E is applied along the chain direction.

The energy related to the spatial and temporal fluctuations of the phase has been derived in Chapter 6. From Eq. (6.7), the Lagrangian density (including the potential energy due to the applied electric field, and neglecting amplitude fluctuations), is given by

$$\mathscr{L} = T - (E_{el} + H_E) = d\left(\frac{d\phi}{dt}\right)^2 - c\left(\frac{d\phi}{dx}\right)^2 - \frac{eEx}{\pi}\frac{d\phi}{dx}. \tag{9.6}$$

The Lagrangian equation

$$\frac{\partial \mathscr{L}}{\partial \phi} - \frac{\partial}{\partial t}\frac{\partial \mathscr{L}}{\partial \phi_t} - \frac{\partial}{\partial x}\frac{\partial \mathscr{L}}{\partial \phi_x} = 0 \tag{9.7}$$

with $\phi_t = \partial\phi/\partial t$ and $\phi_x = \partial\phi/\partial x$ then gives the equation of motion,

$$\frac{d^2\phi}{dt^2} - \hbar v_F^2 \frac{m}{m^*}\frac{d^2\phi}{dx^2} = \frac{2k_F e}{m^*} E(\omega, k) \tag{9.8}$$

in the presence of an *ac* electric field $E(\omega, k) = E_0 e^{i(\omega t - kx)}$. The frequency and wavevector dependent conductivity $\sigma_{\text{coll}}(k, \omega)$ then is given by

$$\sigma_{\text{coll}}(k,\omega) = \frac{j(k,\omega)}{E(k,\omega)} = \frac{m}{m^*}\frac{i\omega\omega_p^2}{\omega^2 - \hbar v_F \frac{m}{m^*} k^2}. \tag{9.9}$$

Because of the short coherence lengths, nonlocal effects are not important and the k dependence can be neglected. Then

$$\sigma_{\text{coll}}(\omega) = \frac{j(\omega)}{E(\omega)} = \frac{m}{m^*} \frac{i\omega_p^2}{4\pi(\omega + i\delta)} \tag{9.10}$$

where $\omega_p^2 = 4\pi ne^2/m$ is the plasma frequency. The real part

$$\text{Re}\{\sigma_{\text{coll}}(\omega)\} = \frac{m}{8m^*}\omega_p^2\delta(\omega) \tag{9.11}$$

has a Dirac delta singularity at $\omega = 0$, with an oscillator strength

$$f = \int \text{Re}\{\sigma_{\text{coll}}(\omega)\}dw = \frac{\pi ne^2}{2m^*}. \tag{9.12}$$

The imaginary part is obtained using the Kramer-Kronig relation

$$\text{Im}\{\sigma_{\text{coll}}(\omega)\} = -\frac{2\omega}{\pi}\int_{-\infty}^{\infty} \frac{\text{Re}\{\sigma_{\text{coll}}(\omega)\}}{\omega'^2 - \omega^2}dw' = \frac{m}{4\pi m^*}\frac{\omega_p^2}{\omega}. \tag{9.13}$$

Because of the gap, the contribution from single particle excitations is expected to be similar to that of a semiconductor with a well defined absorption edge of energy $\hbar\omega = 2\Delta$. In the absence of the contribution from the collective mode, the optical conductivity coming from band-to-band transitions is given by

$$\sigma_{\text{sp}}(\omega) = \frac{ne^2}{i\omega m_b}(f(\omega) - 1) \tag{9.14}$$

where, for a one-dimensional semiconductor, (Wooten 1972)

$$f(\omega) = -\int dE_k \frac{2\Delta^2/E}{(\omega + i\eta)^2 - 4E^2} = \frac{2\Delta^2}{\omega y}\left(\pi i + \ln\frac{1-y}{1+y}\right) \tag{9.15}$$

where $E^2 = E_k^2 + \Delta^2$, $E_k = \epsilon_k - \epsilon_F$, and $y = (1 - 4\Delta^2/\hbar^2\omega^2)^{1/2}$. Because of the $E^{-1/2}$ singularity of the density of states in one dimension, $\sigma_{\text{sp}}(\omega)$ has a singularity at the gap frequency ω_g. The low frequency dielectric constant is

$$\epsilon(\omega \to 0) = 1 - \frac{4\pi\,\text{Im}\{\sigma\}}{\omega} = 1 + \frac{1}{6}\frac{(\hbar\omega_p)^2}{\Delta^2}. \tag{9.16}$$

The collective mode contribution to the conductivity modifies the oscillator strength in (9.14) and is given by (Lee, Rice, and Anderson, 1979)

$$f'(\omega) = \frac{f(\omega)}{1 + \left(\frac{m}{m^*} - 1\right) f(\omega)} = \frac{f(\omega)}{1 + \frac{\lambda \omega_{2k_F}}{4\Delta^2} f(\omega)} \qquad (9.17)$$

and the conductivity reads

$$\sigma(\omega) = \frac{ne^2}{i\omega m}\left[\frac{f(\omega)}{1 + \frac{\lambda \omega_{2k_F}^2}{4\Delta^2} f(\omega)} - 1\right]. \qquad (9.18)$$

For $\omega = 0$, the Kramers-Kronig relation leads to Eq. (9.11) and the frequency dependent response is also modified at energies $\hbar\omega > \Delta$. In Fig. 9.1, $\mathrm{Re}\{\sigma(\omega)\}$ is shown for various values of the effective mass. The mass enhancement, $m^*/m \sim 100$, is a typical value for materials with a CDW ground state (this will lead to a curve similar to that shown in Fig. 9.1); for SDW's the effective mass is expected to be equal to the band mass. This difference leads also to different spectral weights associated with the single particle and collective mode contributions to the optical conductivity.

The relative weights of the collective mode and single particle contributions are determined by sum rule arguments. The total contribution to the sum rule

$$\int \left[\mathrm{Re}\{\sigma_{\mathrm{coll}}(\omega)\} + \mathrm{Re}\{\sigma_{\mathrm{sp}}(\omega)\}\right] d\omega = \frac{\pi n e^2}{2m_b} \qquad (9.19)$$

is the same as the sum rule which is given in the metallic state

$$\int_0^\infty \mathrm{Re}\{\sigma_{\mathrm{m}}\}(\omega)\, d\omega = \int \frac{ne^2\tau}{m_b(1 - \omega^2\tau^2)} d\omega = \frac{\pi n e^2}{2m_b}. \qquad (9.20)$$

Here the integrals include all contributions to σ which come from a band with bandmass m_b. In the density wave state collective mode contribution (see Eq. (9.12)) is

$$I_{\mathrm{coll}} = \int \mathrm{Re}\{\sigma_{\mathrm{coll}}(\omega)\} d\omega = \frac{\pi n e^2}{2m^*} \qquad (9.21)$$

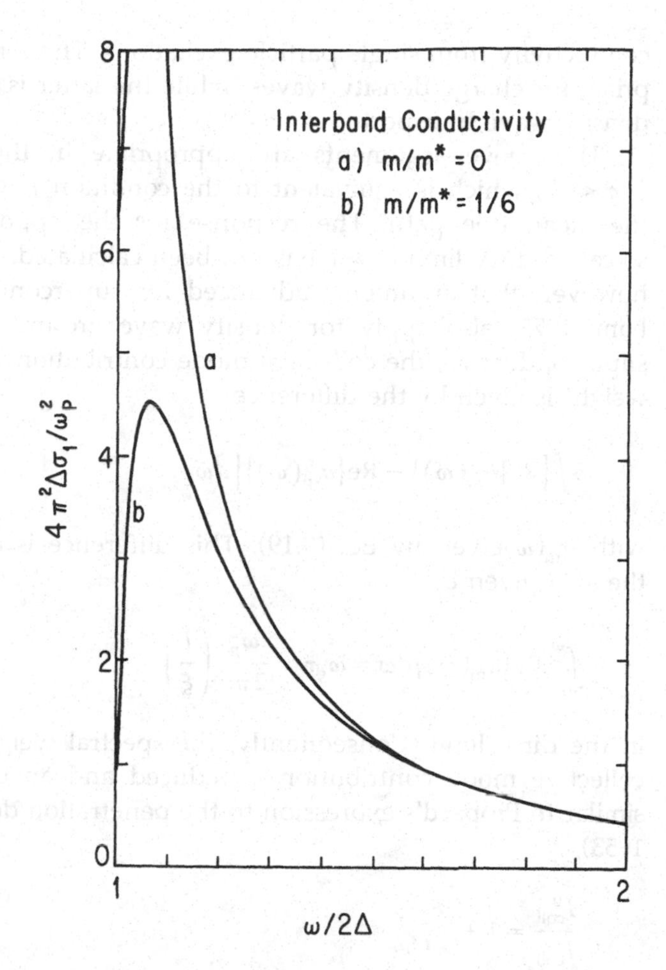

Figure 9.1. The frequency dependence of the conductivity $\mathrm{Re}\{\sigma(\omega)\}$ for $m^*/m = \infty$ and $m^*/m = 6$ (Lee et al., 1974).

while the single particle excitations give the contribution

$$I_{\mathrm{sp}} = \int \mathrm{Re}\{\sigma_{\mathrm{sp}}(\omega)\} d\omega = \frac{\pi n e^2}{2m_b} - \frac{\pi n e^2}{2m^*} = \frac{\pi n e^2}{2m_b}\left[1 - \frac{m_b}{m^*}\right]. \tag{9.22}$$

For a large effective mass, $m^*/m_b \gg 1$, nearly all of the contribution to the total spectral weight come from single particle excitations, while for $m^*/m_b = 1$ all the spectral weight is associated with the collective mode, with no contribution to the optical conductivity from single particle excitations. The former is appropriate for charge density waves, while the latter is valid for spin density wave transport.

The above arguments are appropriate in the clean limit, $1/\tau \ll \Delta$, which is equivalent to the condition $\xi \ll l$ where l is the mean free path. The response for the opposite case, the so-called dirty limit $\xi \gg l$ has not been calculated. It is expected, however, that arguments advanced for superconductors (Tinkham, 1975) also apply for density wave ground states. As for superconductors, the collective mode contribution to the spectral weight is given by the difference

$$\delta \int \left[\mathrm{Re}\{\sigma_{\mathrm{m}}(\omega)\} - \mathrm{Re}\{\sigma_{\mathrm{sp}}(\omega)\}\right] d\omega \tag{9.23}$$

with $\sigma_{\mathrm{m}}(\omega)$given by Eq. (9.19). This difference is approximately the area given by

$$\int_0^{\omega_g} \mathrm{Re}\{\sigma_{\mathrm{m}}(\omega)\} d\omega \simeq \omega_g \tau \simeq \frac{\omega_p^2}{2\pi^2}\left(\frac{l}{\xi}\right) \tag{9.24}$$

in the dirty limit. Consequently, the spectral weight due to the collective mode contribution is reduced and an empirical form similar to Pippard's expression to the penetration depth (Pippard, 1953),

$$\frac{I_{\mathrm{coll}}^0}{I_{\mathrm{coll}}} = 1 + \frac{\xi}{\alpha l}, \tag{9.25}$$

can be anticipated, with α a numerical factor on the order of one. In the above equation I_{coll}^0 is the spectral weight of the collective mode in the clean limit.

For a perfect crystal, and for an incommensurate density wave, the collective mode contribution occurs at $\omega = 0$ due to the translational invariance of the ground state. As has been discussed in Chapter 8, this translational invariance is broken and the collective modes are tied to the underlying lattice due to interactions with impurities. To the first order this can be described by a restoring force $k = \omega_0^2 m^*$. The interaction between the collective mode and the lattice imperfections, impurities, etc., may also lead to a finite relaxation time τ^*. With these effects the equation of motion becomes

$$\frac{d^2\phi}{dt^2} + \frac{1}{\tau^*}\frac{d\phi}{dt} + {\omega_0}^2\phi = \frac{ne}{m^*}E(t) \tag{9.26}$$

leading in the presence of an *ac* electric field $E(t) = E_0(\omega)e^{i\omega t}$, to

$$\mathrm{Re}\{\sigma(\omega)\} = \frac{ne^2\tau^*}{m^*}\frac{\omega^2/\tau^*}{(\omega_0^2-\omega^2)^2+(\omega/\tau^*)^2} \tag{9.27a}$$

$$\mathrm{Im}\{\sigma(\omega)\} = \frac{ne^2}{m^*}\frac{(\omega_0^2-\omega^2)\omega}{(\omega_0^2-\omega^2)^2+(\omega/\tau^*)^2} \tag{9.27b}$$

The collective mode contribution to $\sigma(\omega)$ now appears at finite frequencies, and both Re$\{\sigma(\omega)\}$ and $\epsilon(\omega) = 4\pi Im\sigma(\omega)/\omega$ are shown in Fig. 9.2. There are several characteristic frequencies which are given by the maxima of Re$\{\sigma(\omega)\}$ and the zeros of the dielectric constant Re$\{\epsilon(\omega)\}$. The collective mode contribution, Eq. (9.27), leads to a maximum and to a zero crossing for Re $\sigma(\omega)$ and for Im $\sigma(\omega)$ at ω_0. For $\omega \ll \omega_0$ the dielectric constant is given by

$$\epsilon_0 = 1 + \frac{4\pi ne}{m^*\omega_0^2} + \frac{4\pi ne^2\hbar^2}{m_b\Delta^2} = 1 + \epsilon_{\mathrm{coll}} + \epsilon_{\mathrm{sp}} \tag{9.28}$$

where the second and third terms on the right hand side represent the collective and single particle contributions to the dielectric constant. A second zero crossing occurs at

$$\omega_p^* = \left(\frac{4\pi ne^2}{m^*\epsilon_{\mathrm{sp}}}\right)^{1/2} \tag{9.29}$$

the plasma frequency of the collective mode, and in the region $\omega_p^* < \omega < \Delta/\hbar$ the dielectric constant is

$$\epsilon_{\mathrm{sp}} = 1 + \frac{4\pi ne^2\hbar^2}{m_b\Delta^2}. \tag{9.30}$$

The second maximum of Re$\{\sigma(\omega)\}$ at 2Δ represents the single particle transitions, and Re$\{\epsilon(\omega)\}$ has a crossover from negative to positive values at the plasma frequency $\omega_p = (4\pi ne^2/m_b)^{1/2}$.

For $m^* = m_b$, which is expected for a spin-density wave ground state, single particle excitations do not contribute to the

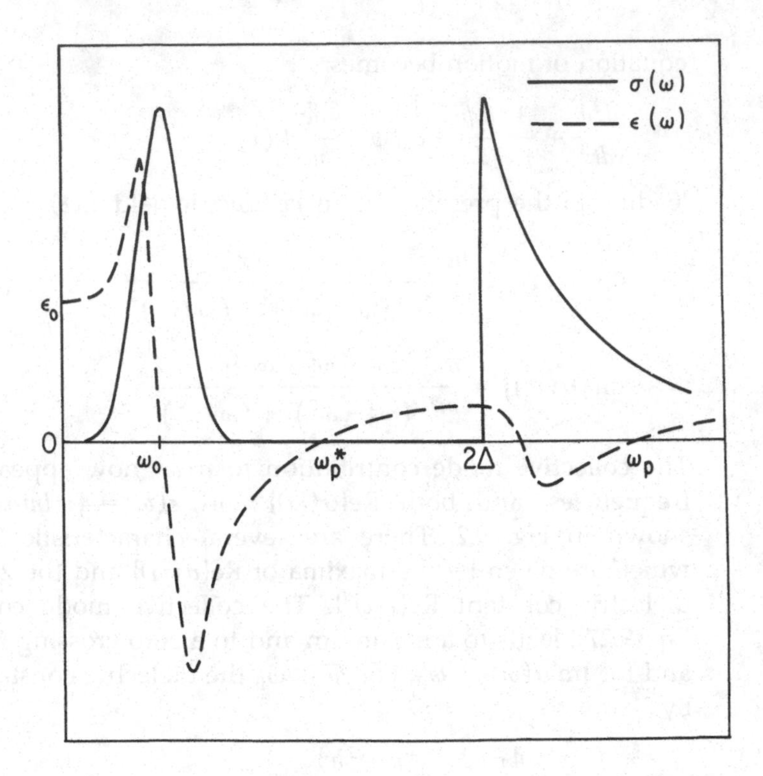

Figure 9.2. The frequency dependence of the optical conductivity $\mathrm{Re}\{\sigma(\omega)\}$ and dielectric constant $\mathrm{Re}\{\epsilon(\omega)\}$ for $m^*/m = \gg 1$. The pinning frequency $\omega_0 < \Delta$, and $1/\tau^* = \omega_0$. The various characteristic frequencies are discussed in the text.

optical conductivity, the collective mode contribution appears at $\omega = \omega_0$, and

$$\epsilon(\omega \to 0) = \frac{4\pi n e^2}{m_b \omega_0^2} \tag{9.31}$$

with a zero crossing at the plasma frequency ω_p.

The effect of impurities described by an average restoring force k is a gross oversimplification since it neglects the dynamics of the local deformations of the collective modes. The types of processes which have been neglected are shown in Fig. 9.3. The top displays an undistorted density wave, with a period $\lambda_0 = \pi/k_F$, and a constant phase ϕ. In the presence of impurities the collective mode is distorted and adjusted to maximize the energy gain due to interaction with the impurity potential; a distorted

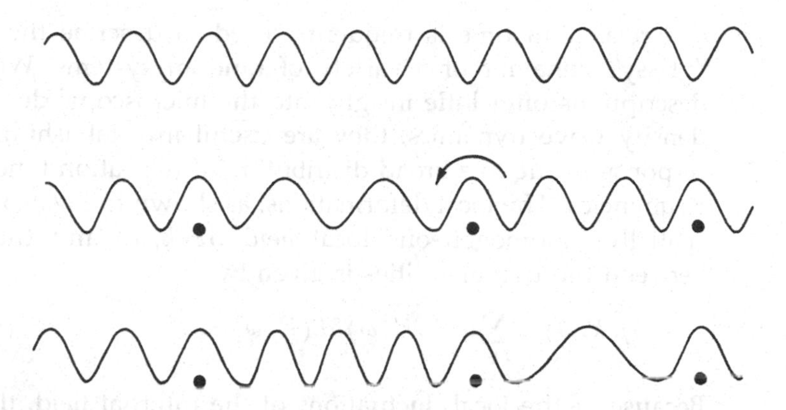

Figure 9.3. The dynamics of the internal deformations of density waves. The top part of the figure is the undistorted density wave; the middle part shows the mode distorted due to interaction with impurities (full circles); and the bottom part displays the rearrangement of the internal distortion by displacing the density wave by period over the impurity as indicated by the arrow. The process leads to an internal polarization.

density wave is shown in the middle section of the figure. A low-lying excitation, which involves the dynamics of the internal deformations, is shown at the bottom of Fig. 9.3; here a density wave segment has been displaced by λ_0, leading to a stretched density wave to the left and to a compressed density wave to the right of the impurity. The local deformation leads to an internal polarization by virtue of the displaced charge which accompanies the stretched or compressed density wave. This polarization is given by

$$P(\vec{r}) = 4\pi e \frac{\partial \phi(\vec{r})}{\partial \vec{r}}. \tag{9.32}$$

Such effects have been described as a broad superposition of Debye-type relaxation processes, and various phenomenological expressions have been proposed to account for the low frequency and long time behavior of the electrical response. Among these, the so-called Cole-Cole expression (see for example, Ngai, 1979 and references cited therein)

$$\epsilon(\omega) = \frac{\epsilon_0}{1 + (i\omega\tau_0)^{1-\alpha}} \tag{9.33}$$

with $\alpha < 1$, and with ϵ_0, and τ_0 an "average" dielectric constant and relaxation time is frequently used to describe the so called "glassy" behavior of a variety of random systems. While these descriptions offer little insight into the microscopic details of the density wave dynamics, they are useful in establishing that the response is due to a broad distribution of relaxation times and/or frequencies. The local deformations, as shown in Fig. 9.3, lead to a spatially inhomogeneous local field $\delta E(k, \omega)$ and the relation between the two quantities is given by

$$\delta j(k, \omega) = \sum_{k'} \sigma(k, k', \omega) \delta E(k', \omega). \tag{9.34}$$

Because of the local fluctuations of the internal field, the experimentally measured conductivity $\sigma(\omega)$ which relates the spatial averages $\delta E(\omega)$ and $\delta j(\omega)$ through the expressions

$$\delta j(\omega) = \sigma(\omega) \delta E(\omega) \tag{9.35}$$

cannot be expressed as $\langle \sigma(k, k', \omega) \rangle = \sigma(k, \omega) \delta_{k,k'}$ and a more sophisticated treatment which includes the k dependence of the electric field fluctuations is required. This complicated issue has received considerable attention recently (Littlewood, 1987; Wonneberger, 1991).

Impurities also lead to changes in the single particle excitation spectrum, in much the same way that impurities lead to bound states in conventional semiconductors (Tutto and Zawadowski, 1985; Zawadowski, 1987.

9.2 Frequency Dependent Conductivity of Charge Density Waves

The electrical conductivity has been investigated over a broad spectral range in various materials and $\mathrm{Re}\, \sigma(\omega)$ is shown in Fig. 9.4 (Grüner, 1988) for some of the compounds in their charge density wave state. For clarity, phonon modes observed in the far infrared spatial range are omitted from the figure. The strong rise at frequencies approximately $10^{14} \mathrm{sec}^{-1}$ is due to single particle excitations across the gap. The arrows on the figures indicate the gap values which are obtained from the temperature dependence of the low field *dc* conductivity in the charge density wave state. The resonance which is observed in the microwave spectral range is due to the pinned collective mode with the resonance shifted to finite frequencies due to the interaction with impurities. This has

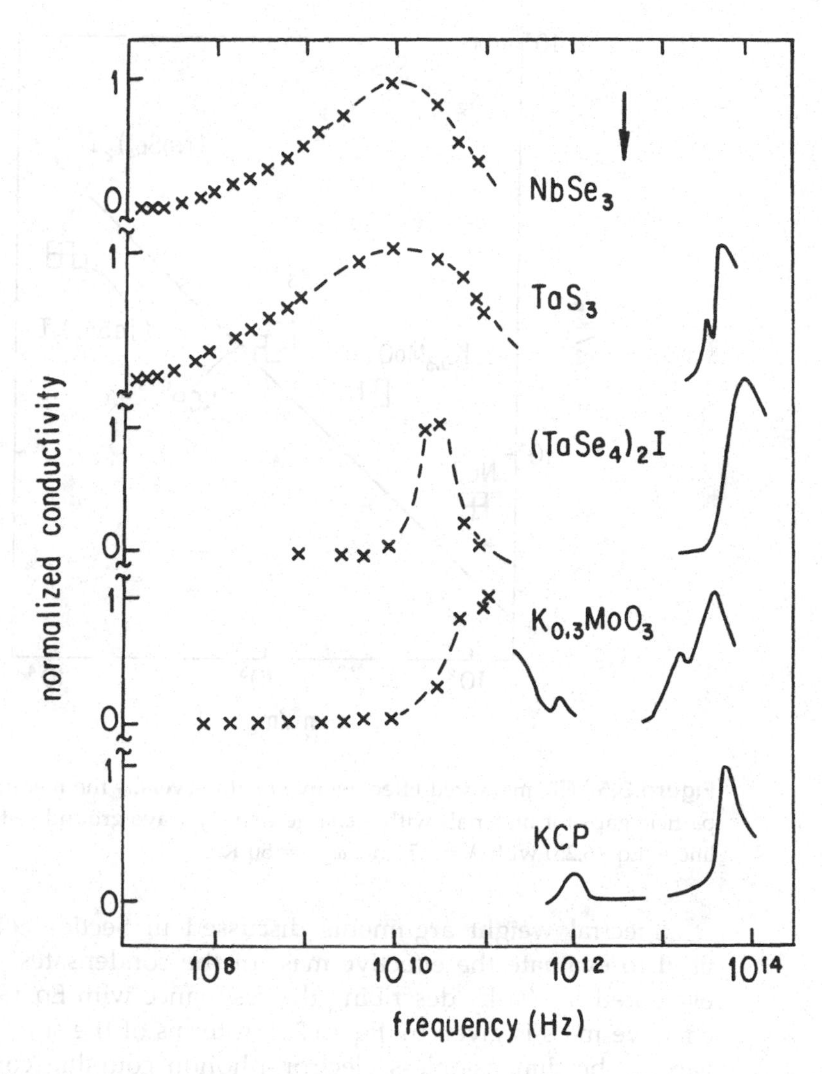

Figure 9.4. Frequency dependent conductivity measured in several compounds in their charge density wave state (Grüner, 1988). The arrow indicates the gap measured by tunneling for $NbSe_3$.

been directly confirmed by frequency dependent conductivity studies conducted in alloys (Kim et al., 1991) where increasing impurity concentrations lead to increasing pinning frequencies ω_0. The finite width of the collective mode resonances is due to damping effects, which to date have not been addressed by theory.

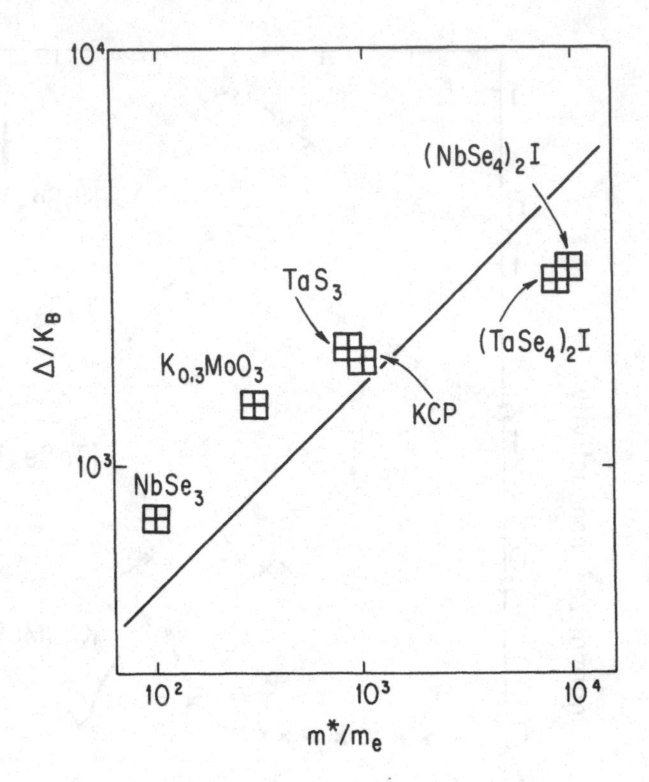

Figure 9.5. The measured effective mass values versus the measured single particle gaps for materials with a charge density wave ground state. The full line is Eq. (6.25) with $\lambda = 0.5$ and $\omega_{2k_F} = 50$ K.

Spectral weight arguments discussed in Section 9.1 can be used to evaluate the effective mass of the condensates. We have evaluated m^*/m by describing the resonance with Eq. (9.27). The effective mass is given by Eq. (6.25) in terms of the single particle gap Δ, the dimensionless electron-phonon coupling constant λ, and the unrenormalized phonon frequency $\omega^0_{2k_F}$. For the materials investigated, λ is somewhat less than one, while $\omega^0_{2k_F}$ is on the order of 5×10^{-2} eV and the variation of the mass enhancement from material to material is mainly due to differences in the magnitude of the single particle gaps. This is clearly shown in Fig 9.5 where the effective mass values are plotted as functions of the measured gap values. The full line on the figure is Eq. (6.25) with $\lambda = 0.5$ and ω_{2k_F} as given above.

The single particle contribution to the optical conductivity is described by Eq. (9.18); the expression leads to a square root

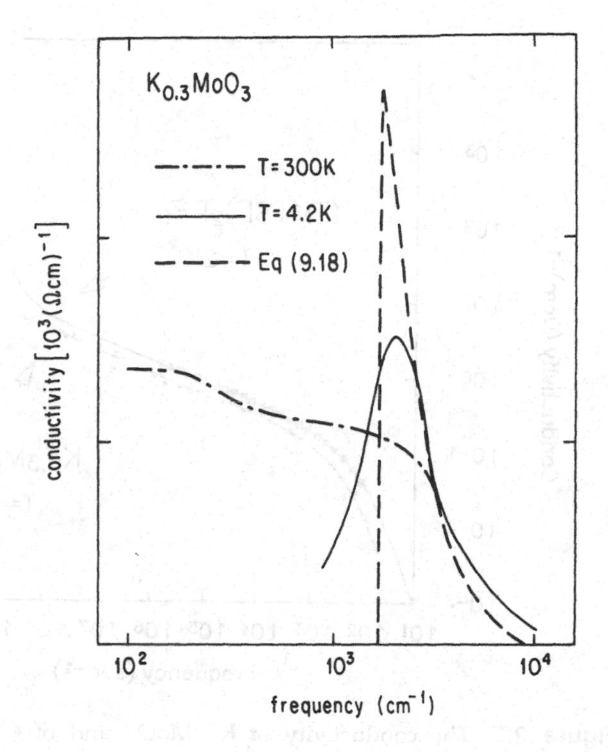

Figure 9.6. Measured and calculated optical conductivity for $K_{0.3}MoO_3$ in the charge density wave state. The conductivity measured above the charge density wave transition is also shown.

divergence of Re $\sigma(\omega)$ for large effective mass at $\omega_g = 2\Delta/\hbar$. This feature is clearly evident from the data shown in Fig. 9.4. The optical conductivity measured on $K_{0.3}MoO_3$ (Degiorgi et al., 1991) is compared with $\sigma(\omega)$ as given by Eq. (9.18) in Fig. 9.6, where the conductivity measured above the CDW transition is also shown. Several aspects of the single particle response are evident on the figure. The spectral weight of the single particle excitations at $T = 4.2$ K is (within the experimental error) the same as the spectral weight associated with the Drude response above the transition. As expected, due to the large effective mass the contribution of the collective mode to the total spectral weight is small (see Eqs. (9.21) and (9.22)). The measured conductivity $\sigma(\omega)$ also has a well defined peak reminiscent of the singularity which is given by Eq. (9.15). The broadened peak (in contrast to the sharp feature expected) may reflect the anisotropy of the single particle

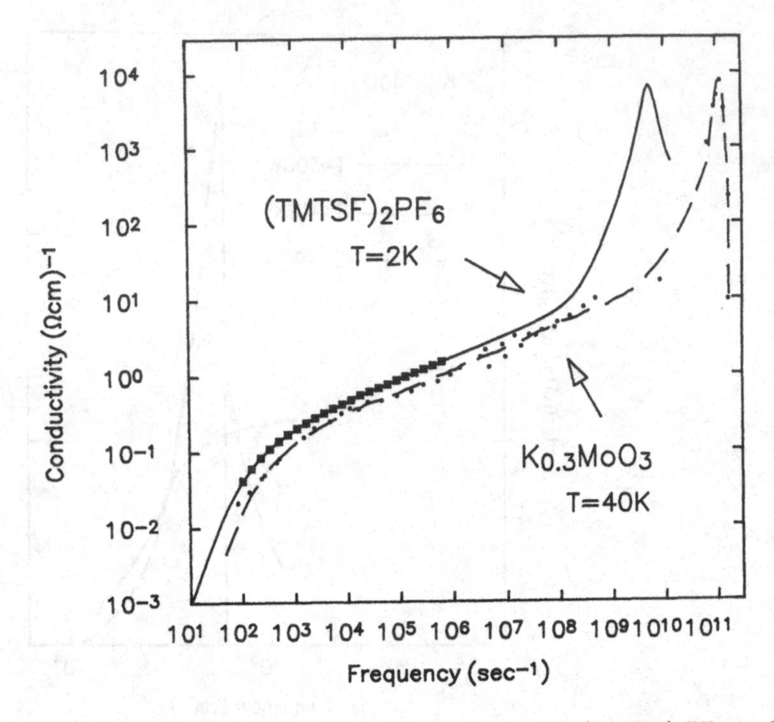

Figure 9.7. The conductivity of $K_{0.3}MoO_3$ and of $(TMTSF)_2PF_6$ at low frequencies. The broad tail is due to the internal modes in both materials. The full and dotted lines are a guide to the eye.

gap but may also be due to impurities which are expected to remove the singularity and also introduce states in the gap.

The low frequency response has also been investigated in detail and the behavior is displayed in Fig. 9.7 in one of the compounds with a CDW ground state, $K_{0.3}MoO_3$ (Degiorgi et al., 1991). The resonance which occurs at $\omega_0/2\pi = 100$ GHz is due to the oscillations of the collective mode around the pinned equilibrium position, and to a first approximation, the resonance can be described by Eq. (9.27). The low frequency tail of the spectrum is due to the dynamics of the internal deformations of the collective mode, and the behavior has been described by the Cole-Cole expression, Eq. (9.33), or its modified variants. A theory which takes the local fluctuations of the condensate into account (Wonneberger, 1991 and references cited therein), can account for the collective mode dynamics in detail, with only a few free parameters.

The previous discussion of the single particle and collective mode contributions to the conductivity neglected the interaction between the ground state and optical phonons which occurs at frequencies ω_n. Such an interaction has several important consequences and leads to modifications of the phonon contributions to $\sigma(\omega)$ both as far as their spectral intensity and polarization are concerned. The reason for this is that the oscillation of the ionic positions at frequency ω_m leads, through a coupling to the charge density wave modulation, to oscillations of the condensate. Such oscillations can be described by the oscillations of the phase, with frequency ω_n. The polarization of the mode is then parallel to the wavevector of the charge density wave. For a spectrum of optical phonons at frequencies ω_n and coupled to the CDW with a dimensionless coupling constant λ_n, the optical conductivity is given by (Rice, 1976)

$$\sigma(\omega) = \frac{ne^2}{i\omega m_b}\left[f(\omega) - f(0) - \left(\frac{\hbar\omega}{2\Delta}\right)^2 f^2(\omega)\lambda' D_\phi(\omega)\right]. \tag{9.36}$$

The first two terms describe the optical conductivity of a one-dimensional semiconductor. The last term is similar to the renormalization of $f(\omega)$, as discussed earlier for the zero frequency collective mode. Here however, $D_\phi(\omega)$ is due to coupling to the phonon modes. The propagator $D_\phi(\omega)$ describes the set of optical phason modes and this, by virtue of Eq. (9.36), leads to maxima in $\text{Re}\{\sigma(\omega)\}$ at frequencies close to ω_n. The spectral weight of these modes is determined by the coupling constants λ_n, and due to the conservation of the total spectral weight, the spectral weight of the single particle excitations is reduced. The theory was first applied (Rice, 1976) to the CDW state of the organic conductor $TEA(TCNQ)_2$, and recently to the charge density wave state of $K_{0.3}MoO_3$ (Degiorgi et al., 1991).

9.3 Frequency Dependent Conductivity of Spin Density Waves

The frequency dependent response of spin density waves is expected to be fundamentally different from the frequency dependent conductivity observed in the charge density wave state. Because the effective mass m^* is the same as the band mass, all

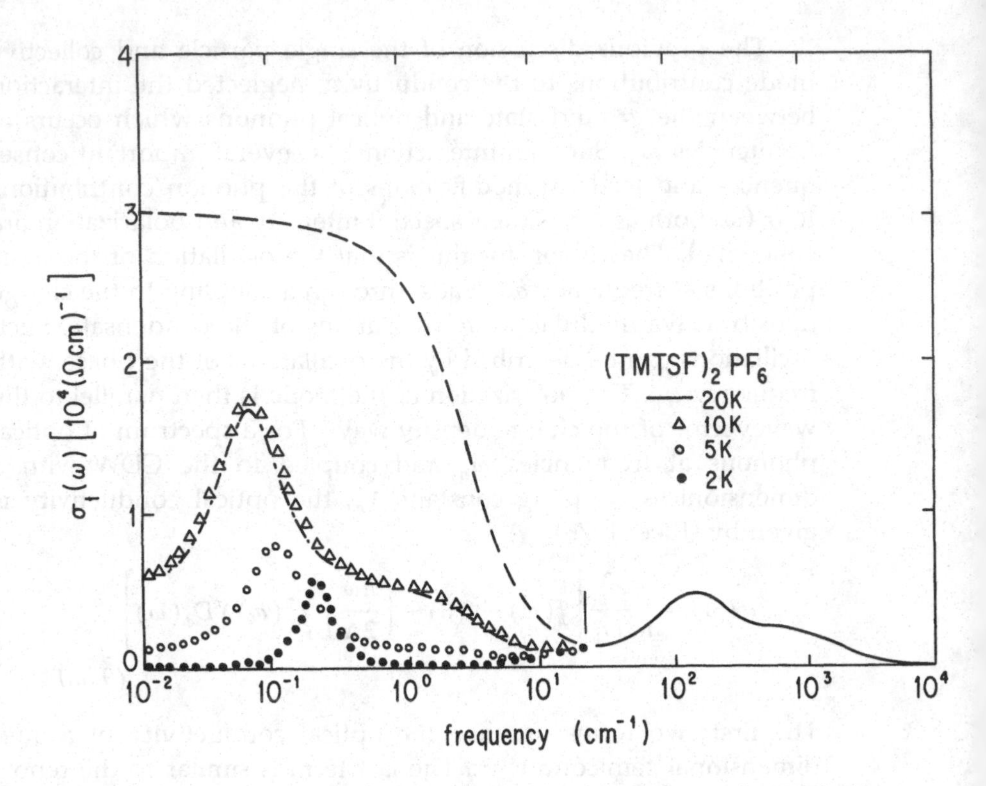

Figure 9.8. The frequency dependent conductivity in $(TMTSF)_2PF_6$ at several temperatures. Above $T_{SDW} = 11.5$ K the low frequency conductivity is that of a Drude metal; below T_{SDW} the low frequency peak is due to the pinned mode.

the spectral weight is expected to be associated with the collective mode, with single particle excitations not contributing to $\sigma(\omega)$ in the clean limit; this limit is appropriate in $(TMTSF)_2X$ salts which have been investigated in detail. The frequency dependent conductivity measured in $(TMTSF)_2PF_6$ (Quinlivan et al., 1991; Donovan et al., 1992), shown in Fig. 9.8; is in clear disagreement with the models of spin density wave dynamics discussed in Section 9.1. As for charge density waves, we recover a resonance in the microwave spectral range; the most likely explanation of which is that this resonance is the pinned spin density wave mode. The pinning frequency, $\omega_0 \simeq 3$ GHz, is smaller than the single particle gap Δ. The increase of $\mathrm{Re}\{\sigma(\omega)\}$ in the far infrared spectral range is not due to single particle excitations; the feature has also been observed above the transition temperature at 11.5 K. The resonance which occurs at frequencies near $w_0/2\pi = 3$ GHz has a small weakly temperature dependent spectral weight (Donovan et al., 1992) and is approximately two orders of magnitude smaller than the spectral weight which would correspond to an effective mass equal to the band mass m_b. This gross disagreement between theory and experiment is unaccounted for at present.

The dynamics of the internal deformations of the collective mode have also been examined and, as for charge density waves, a low frequency tail develops as a consequence of these deformations. This is shown in Fig. 9.7 where $\mathrm{Re}\{\sigma(\omega)\}$ measured in the spin density wave state of $(TMTSF)_2PF_6$ is compared with $\mathrm{Re}\{\sigma(\omega)\}$ measured in the charge density wave state of $K_{0.3}MoO_3$. Theories derived for the low frequency dynamics of charge density wave states (Littlewood, 1987; Wonneberger, 1991) are also expected to give a good account of the experimental state for the spin density wave response.

Nonlinear Transport

de mégis utnak indul, mint akit szárny emel
s hiaba hivja árok, maradni ugyse mer

as if picked up by wings, moving along again
and though the ditch is calling, dares not long to remain
—Miklos Radnoti *Forced March*

10

In the charge and spin density state the *dc* electrical conductivity σ_{dc} is that of a semiconductor, with the temperature dependence determined by carriers excited across the (temperature dependent) single-particle gap. As discussed in the previous chapter, electrons which form the charge or spin density wave condensate do not participate in the conduction process for small *dc* electric fields. The reason for this is the interaction between the condensate and impurities and other lattice irregularities. This interaction shifts the oscillator strength associated with the collective modes to finite frequencies, with no contribution to the *dc* conductivity. The small characteristic frequencies which describe the oscillations of the collective mode about the pinned equilibrium positions are suggestive of weak restoring forces. It can be anticipated that small *dc* electric fields applied along the direction where the incommensurate density wave is developed can therefore induce a translational notion of the condensate; this phenomenon is called sliding density wave transport.

As was discussed previously, the dynamics of internal deformations play an important role in the frequency dependent response. It is therefore expected that such effects are also important for the *dc* nonlinear dynamics of the driven collective modes.

However, a highly oversimplified model—essentially an extension of the harmonic oscillator approach used to describe the *ac* response—reproduces many of the essential features of the experimental findings on the nonlinear transport process. As expected, the model is not capable of accounting for many observations which are related to the dynamics of the internal deformations. These have been treated by various approaches (see, for example, Littlewood, 1989 and references cited therein).

There is a vast amount of literature on the subject, which has been summarized in several reviews listed in the Appendix. We also find a broad variety of nonlinear effects, depending on the nature of the pinning boundaries (whether pinning is due to a dense set of weak impurity potentials or to a single grain boundary, for example), the size of the specimens (which can be larger or smaller than the phase-phase correlation length L_0 discussed in the previous chapter), or temperature—just to mention a few important variables. Therefore it is expected that no single model will account for the variety of observations.

10.1 Models of Density Wave Transport

A simple description of nonlinear transport due to moving density wave condensates neglects the internal dynamics of the mode and treats the dynamics of the rigid condensate, neglecting finite temperature fluctuation effects. It is assumed that the overall effect of the impurities can be described by an impurity potential V which pins the density wave to the underlying lattice; this then leads to an equilibrium density wave position at $x = 0$ where the density wave rests at the bottom of the potential well. Because of the intrinsic periodicity of the charge or spin density wave, the interaction energy which results from the interaction between impurities and the collective mode is written as

$$\mathscr{H} = V \cos(2k_F x + \phi) \tag{10.1}$$

where the interaction potential is a simple periodic function of the displacement x. Neglecting quantum and finite temperature effects, the problem is then that of a classical particle moving in a one-dimensional periodic potential, which is assumed to have the

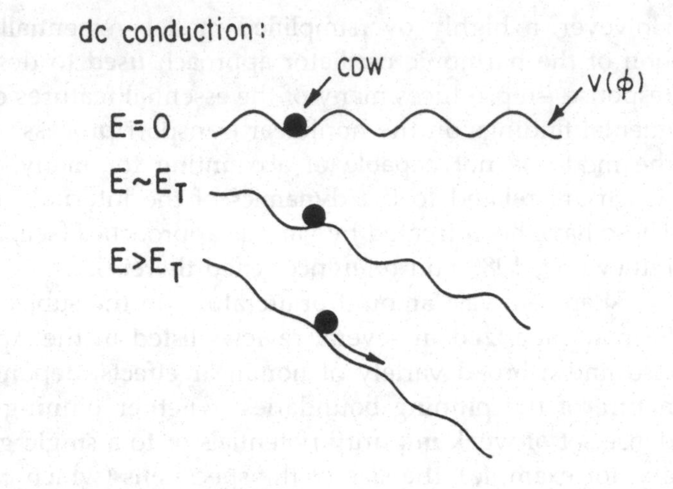

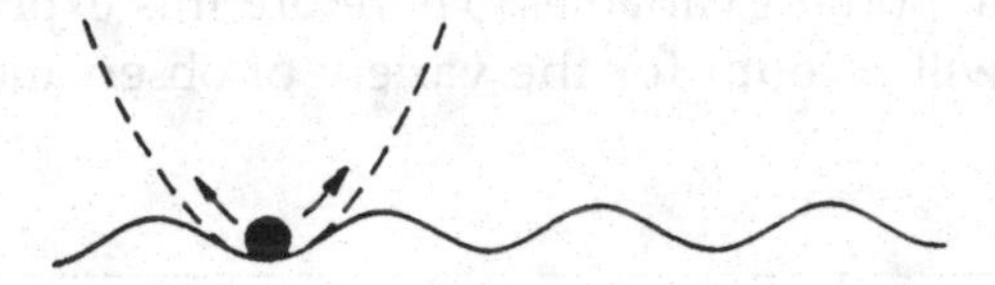

Figure 10.1. The classical particle model of density wave transport. The upper part of the figure indicates the behavior in the presence of a *dc* electric field, the lower part refers to the response to small amplitude *ac* excitations.

form

$$V(x) = \frac{\omega_0}{2k_F} \sin 2k_F x. \tag{10.2}$$

The driving force due to the applied electric field is $F = eE$, and the equation of motion of the collective coordinate is given by

$$\frac{dx^2}{dt^2} + \frac{1}{\tau}\frac{dx}{dt} + \frac{\omega_0^2}{2k_F} \sin 2k_F x = \frac{eE}{m^*}. \tag{10.3}$$

The equation of motion for small amplitude displacements reduces to the equation (9.26). The consequences of this so-called classical particle model (Grüner et al., 1981) is shown for various *dc* electric fields in Fig. 10.1. For small *dc* fields the particle resides in the bottom of the slightly tilted well; there is no *dc* current which would correspond to the translational motion. There is however a finite displacement which corresponds to a finite polarization P. With the increase of the *dc* field there is a well defined threshold field E_T, where the particle starts to roll down in the staircase potential. The condition for this is

$$eE_T\lambda_0 = \frac{2m^*\omega_0^2}{2k_F} \tag{10.4}$$

from which the threshold field is given by

$$E_T = \frac{m^*\omega_0^2}{2\pi ne}. \tag{10.5}$$

For small velocities the inertial term can be neglected and the equation of motion reads

$$\frac{1}{\tau}\frac{dx}{dt} = \frac{eE}{m^*} - \frac{\omega_0^2}{2k_F}\sin 2k_F x. \tag{10.6}$$

The time dependent displacement is given by the equation

$$\frac{1}{\tau}\int_0^{x(t)} \frac{dx}{\frac{eE}{m^*} - \frac{\omega_0^2}{2k_F}\sin 2k_F x} = \int_0^t dt' = t. \tag{10.7}$$

The integral can be evaluated and the time dependent displacement is determined by the equation

$$\frac{2}{\left[\left(\frac{eE}{m^*}\right)^2 - \left(\frac{\omega_0^2}{2k_F}\right)^2\right]^{1/2}} \tan^{-1} \frac{\frac{eE}{m^*} - \left(\frac{\omega_0^2}{2k_F}\right)^2 \tan\frac{x}{2}}{\frac{eE}{m^*} - \frac{\omega_0^2}{2k_F}} = t\tau. \tag{10.8}$$

In the limit of large electric fields $E \gg E_T$, $eE/m^* \gg \omega_0^2/2k_F$ and therefore

$$\frac{2}{\frac{eE}{m^*}} \tan^{-1}\left\{\tan\frac{x}{2}\right\} = t\tau \tag{10.9}$$

which gives

$$x(t) = \frac{eE\tau}{m^*}t \tag{10.10}$$

and consequently a *dc* current

$$I = ne\frac{dx(t)}{dt} = \frac{ne^2\tau}{m^*}E. \tag{10.11}$$

The *dc* conductivity

$$\sigma = \frac{ne^2\tau}{m^*} \tag{10.12}$$

is independent of electric field in this high field limit, it's value is determined by damping effects. At intermediate electric fields Eq. (10.8) represents a time dependent displacement, resulting in a time-dependent current. The time average is given by

$$\langle \tan^{-1}(y \tan y)\rangle_t = y \tag{10.13}$$

and consequently

$$\frac{2}{\left[\left(\frac{eE}{m^*}\right)^2 - \left(\frac{\omega_0^2}{2k_F}\right)^2\right]^{1/2}}\left\langle\frac{x}{2}\right\rangle_t = \tau t \tag{10.14}$$

where $\langle x\rangle$ refers to the time averaged, *dc* displacement

$$\langle x\rangle = \frac{e\tau}{m^*}\left(E^2 - E_T^2\right)^{1/2}. \tag{10.15}$$

The current

$$\langle j\rangle = ne\frac{d\langle x\rangle}{dt} = \frac{ne^2\tau}{m^*}\left(E^2 - E_T^2\right)^{1/2} \tag{10.16}$$

is a strongly nonlinear function of the applied *dc* electric field. Because of this nonlinearity, two conductivities can be defined

$$\sigma_c = \frac{\langle j\rangle}{E} = \frac{ne^2\tau}{m^*}\left[1 - \left(\frac{E_T}{E}\right)^2\right]^{1/2} \tag{10.17}$$

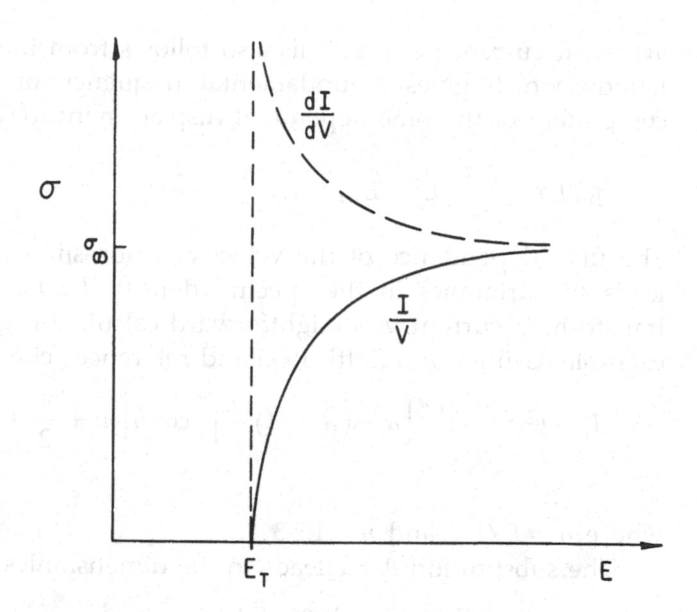

Figure 10.2. Cordial ($\sigma_c = \langle j \rangle / E$) and differential ($\sigma_d = d\langle j \rangle / dE$) conductivity which follows from the classical particle model.

the cordial, and

$$\sigma_d = \frac{d\langle j \rangle}{dE} = \frac{ne^2}{m^*}\left[1 - \left(\frac{E_T}{E}\right)^2\right]^{-1/2} \tag{10.18}$$

the differential conductivity. Both are shown in Fig. 10.2 as the function of applied *dc* electric field.

The equation of motion also leads to a time-dependent current, due to the acceleration and deceleration as the particle moves down in the staircase potential. The fundamental frequency corresponds to the displacement of the particle by one period $\lambda_0 = \pi / k_F = 2/n$ with n the number of electrons per unit length. The time average velocity is $\langle V \rangle = \lambda_0 f_0$, and the time averaged current density per chain is given by

$$\langle j \rangle = ne\langle V \rangle = ne\lambda_0 f_0, \tag{10.19}$$

and consequently,

$$\langle j \rangle / f_0 = 2e \tag{10.20}$$

i.e., the frequency of the oscillations is proportional to the time averaged current density. This also follows from the equation of motion which gives a fundamental frequency of the periodic component of the time dependent displacement $x(t)$,

$$f_0(E) = \frac{e^2\tau}{m^*}(E^2 - E_T)^2. \tag{10.21}$$

The time dependence of the velocity is non-sinusoidal and this leads to harmonics in the spectral density $I(\omega)$ of the Fourier transformed current. A straightforward calculation gives (see, for example, Grüner and Zettl, 1985 and references cited therein)

$$I_n \sim (\alpha^2 - 1)^{1/2}\left[\alpha - (\alpha^2 - 1)^{1/2}\right]^n \cos n\left[nt + \frac{\pi}{2} + \sin^{-1}\alpha\right] \tag{10.22}$$

where $\alpha = E/E_T$, and $n = 1,2,3\ldots$.

The substitution $\theta = x$ leads to the dimensionless form

$$d^2\theta/dt^2 + \Gamma\, d\theta/dt + \sin\theta = E/E_T \tag{10.23}$$

where $\Gamma = (\omega_0\tau)^{-1}$, $E_T = (\lambda_0/2\pi)(m^*\omega_0^2/e)$, and time t is measured in units of ω_0^{-1}. Equation (10.23) is formally identical to the equation which describes the behavior of resistive shunted Josephson junctions (see, for example, Barone and Paterno, 1982)

$$\frac{d^2\phi}{dt^2} + G\frac{d\phi}{dt} + \sin\Phi = I/I_J \tag{10.24}$$

where Φ is the phase difference across the junction, I is the current through the junction; and $G = (RC\omega_J)^{-1}$, where R and C are the resistance and capacitance of the junction, and $\omega_J = 2eI_J/Cg$. I_J is the *dc* critical Josephson current, and time is measured in units of ω_J^{-1}. This analogy suggests a close formal correspondence between phenomena observed in Josephson junctions and in materials displaying CDW transport. For example, the current oscillations described by Eq. (10.20) correspond to the *ac* Josephson effect while the threshold field E_T corresponds to the critical current I_J. This analogy has been extensively used to analyze a variety of interference experiments in the presence of combined *dc* and *ac* electric fields (Grüner, 1988 and references cited therein). Equation (10.23) is also analogous to the equation describing other systems; for example, a damped pendulum in a gravitational field or a phase-locked loop configuration of a voltage controlled oscillator. The equations of motion for these different systems are summarized in Table 10.1.

The model, which neglects the dynamics of the internal deformations of the collective mode, is expected to be appropriate only under circumstances where these deformations play a secondary role; this may happen in specimens with dimensions comparable to or smaller than the phase-phase correlation length discussed in Chapter 8. Since in the majority of specimens investigated this correlation length is of the order of microns for pure specimens, a full treatment of the dynamics of internal modes is required.

In the first approximation we assume that the amplitude of the density wave is constant, and only the phase is locally distorted around the impurities. Then interaction between the collective mode and impurities is given by Eq. (7.1); the full equation of motion, which takes this interaction into consideration reads

$$m^* \frac{d^2\phi(\vec{r})}{dt^2} + \Gamma \frac{d\phi(\vec{r})}{dt} \sum_i \rho_1 V_0\left(\vec{r} - \vec{R}_i\right) \sin\left(2k_F \vec{r} + \phi(\vec{r})\right)$$
$$= 2k_F eE \qquad (10.25)$$

where R_i (describing the impurity positions) is a random variable. The treatment of the nonlinear dynamics of this driven system is complex, and various approaches have been taken to discuss the nonlinear response in certain limits; they have been the subject of several reviews (Littlewood 1990, and references cited therein). Due to this complexity, computer simulations have been extensively employed and compared with the experimental results. An alternate approach (see, for example, Narayan and Fisher, 1992; Myers and Sethna, 1993) treats the problem of depinning as a critical phenomenon which leads to scaling forms for the response and correlation functions near the threshold field E_T. The exponents which determine the electric field dependence of the various parameters are somewhat different from those which follow from the single particle model.

This model of density wave transport applies to a situation where pinning by weak, randomly distributed impurities plays an important role. Under these circumstances the amplitude of the collective mode is not influenced by the impurity potentials. The dynamics are governed by the dynamics of the phase variable

Table 10.1. Correspondence between the motion of pendulum under gravitation, the Josephson equation and the classical model of CDW Transport.

CDW	*Josephson Junctions*	*Pendulum*
applied field E	*dc* current I_{dc}	applied torque τ_a
mass m^*	capacitance C	moment of inertia M_1
damping coeff. $1/\tau$	conductance $1/R$	damping coeff. D_f
threshold field E_T	maximum Josephson current I_j	maximum torque due to gravity mgl
periodic potential Qx	phase difference ϕ	angle from vertical θ
$\frac{d^2x}{dt^2} = \frac{1}{\tau}\frac{dx}{dt} + \frac{\omega_0^2}{\theta}\sin\theta x = \frac{eE}{m}$	$I_{cd} = \frac{h}{2e}C\frac{d^2\phi}{dt^2} + \frac{h}{2e}\frac{1}{R}\frac{d\phi}{dt} + I\sin\phi$	$\tau_a = M_1\frac{d^2\theta}{dt^2} + D_f\frac{d\theta}{dt} + mgl\sin\theta$
$\frac{d^2\theta}{dt^2} + \Gamma\frac{d\theta}{dt} + \sin\theta = \frac{E}{E_T}$	$\frac{d^2\phi}{dt^2} + G\frac{d\phi}{dt} + \sin\phi = \frac{I}{I_J}$	

$\phi(\vec{r})$. In the case of macroscopic defects, such as dislocations or grain boundaries (or even the boundaries associated with contacts), a different model should apply. At such presumably strong pinning centers the translational motion of the condensate under the influence of the *dc* electric field may lead to a transient, periodic collapse of the amplitude of the density wave. This in turn, leads to a 2π phase slip which relieves the elastic strain that accumulates due to the applied electric field. The Ginzburg-Landau expansion which also includes the amplitude fluctuations is given by (Gor'kov 1977)

$$\mathscr{H} = \mathscr{H}_{el} + \mathscr{H}_{\Delta} + \mathscr{H}_{E} + V_{\text{imp}}(\phi). \tag{10.26}$$

The energy due to the applied electric field, $\mathscr{H}_E$ is given by Eq. (9.5). The gradient energy is given by

$$\mathscr{H}_{el} = \int dr \left\{ C_{\Delta}(\nabla|\Delta|)^2 + C_{\phi}(\nabla\phi)^2 \right\} \tag{10.27}$$

where C_{Δ} and C_{ϕ} refer to the elastic constants, related to the change of the amplitude and phase of the order parameter. The energy associated with small amplitude fluctuations by

$$\mathscr{H}_{\Delta} = a \int dr (|\Delta| - |\overline{\Delta}|)^2 \tag{10.28}$$

where $|\overline{\Delta}|$ is the time average order parameter, and a is a constant. The impurity term is given by Eqs. (8.2) or (8.3).

The equation of motion, schematically written as

$$\frac{1}{\tau}\frac{d\Delta}{dt} = -\frac{\delta\mathscr{H}}{\delta\Delta} \tag{10.29}$$

is complicated, even for rather simple potentials, such as the plane boundary at a current or voltage contact. In this case, the strain accumulated at the boundary due to the applied electric field is periodically relieved due to phase slip processes; with every phase slip leading to an advancement of the density wave by the period λ_0. The role of phase slips, and consequently the current carried by the condensate, is determined by the elastic constants C_{Δ} and C_{ϕ}, by the relaxation time τ, and by the applied electric field E. The model also leads to a well defined threshold field where phase slips start to occur; current oscillations and nonlinear current-voltage characteristics are also a natural consequence of the equation of motion.

10.2 Experiments on the Nonlinear Dynamics of the Collective Modes

A broad variety of experiments have been conducted to establish the essential features of the nonlinear conduction caused by the driven collective modes. These include measurement of the *dc* current-voltage characteristics, response to the system to combined *dc* and *ac* electric fields and evaluation of the spectral response, and fluctuations of the nonlinear current. Several review papers and Conference Proceedings listed at the end of the book in the Appendix deal with this subject, and therefore only a short summary will be given here.

Depending on factors like the size of the specimens, impurity concentration, and the presence (or absence) of extended defects such as grain boundaries; a variety of phenomena have been observed. The reason for this is that the dynamics of the collective mode depend sensitively on the nature of pinning provided by the various potentials $V(\vec{r})$ which break the translational invariance of the collective modes. In small specimens, with dimensions comparable to the phase-phase correlation length L_0, the dynamics of the internal deformations do not play a significant role; and the classical particle model is a good first approximation. For such sample dimensions, pinning by the surface of the specimens may also be important. In specimens where pinning is dominated by weak impurities, the dynamics are governed by Eq. (10.25) and the full treatment of internal deformations is required. In materials where pinning is principally due to one strong pinning center, the dynamics of both amplitude and phase fluctuations must be considered with phase slips and amplitude collapse at the pinning site dominating the dynamics of the collective mode. These different conditions lead to different current voltage characteristics, which are also influenced by the (temperature dependent) screening provided by the normal carriers which are excited across the single particle gap. Not surprisingly, therefore, a broad variety of observations have been studied under different conditions.

The *dc* conductivity, defined as j/E where j is the total current density and E is the applied electric field, is shown in orthorhombic TaS_3 in Fig. 10.3. The observation is typical of that found below T_{CDW} in relatively large specimens. Below a thresh-

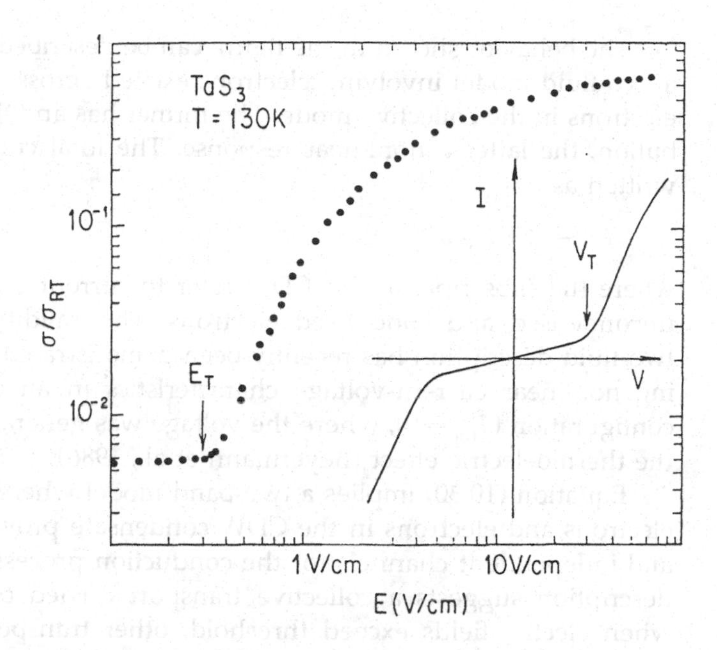

Figure 10.3. Electric-field-dependent cordial conductivity $\sigma_c(E)$ in $o-\mathrm{TaS_3}$ in the CDW state. The data are normalized to the room-temperature conductivity. The insert shows typical *dc* $I-V$ characteristics on the same material.

old field E_T, approximately 300 mV/cm, the conductivity obeys Ohm's law, and the temperature dependence of this component reflects the exponential freezing out of the electrons excited across the single-particle gap. The onset of nonlinear conduction at E_T is smooth, as evidenced by the current-voltage characteristics displayed in the figure. A behavior similar to that displayed in Fig. 10.3 is found, in general, for the other compounds (first on $\mathrm{NbSe_3}$, Monceau et al., 1976); the existence of the sharp threshold field E_T is well established.

The threshold field E_T increases with increasing impurity concentrations, and experiments on various materials either lead to $E_T \sim n_i$ or $E_T \sim n_i^2$. The former implies strong, the latter weak, impurity pinning (see Eqs. (8.8) and (8.15)). The issue is complicated; since factors such as the size of the specimens, dimensionality effect, or other defects (grain boundaries, for example) are also important in determining the overall restoring force.

The behavior shown in the figure can be described in terms of a two-fluid model involving electrons excited across the gap and electrons in the collective mode. The former has an Ohmic contribution, the latter a nonlinear response. The total current can be written as

$$I_{\mathrm{tot}} = I_n + I_{\mathrm{CDW}} \tag{10.30}$$

where the subscripts n and CDW refer to current carried by the uncondensed and condensed electrons. The validity of such a two-fluid description has recently been demonstrated by generating nonlinear current-voltage characteristics in an open circuit configuration ($I_{\mathrm{tot}} = 0$), where the voltage was generated through the thermoelectric effect (Beyermann et al., 1986).

Equation (10.30) implies a two-band model where the normal electrons and electrons in the CDW condensate provide separate and independent channels for the conduction process. While this description suggests a collective transport carried by the CDW when electric fields exceed threshold, other transport measurements performed in the nonlinear conductivity region provide direct evidence that the current at high fields is carried by a ground-state condensate (Grüner, 1988). The onset of nonlinear *dc* transport also leads to changes in other transport coefficients such as Hall effect and thermoelectric power. These have been analyzed in terms of the Onsager relations, which were extended for the nonlinear transport region. Although changes in various thermodynamic properties (such as the elastic contents) have been investigated, a microscopic theory of these effects is not available to date.

The fact that the charge-density wave contributes to the *dc* current for $E > E_T$ rather than being destroyed, has been established by performing X-ray experiments in the nonlinear conduction region (Fleming et al., 1978). The superlattice reflections were found to be independent of the applied electric field, even well into the nonlinear conduction region. The observation demonstrates that the charge-density wave amplitude remains unchanged, and the condensate executes a translational motion in the presence of an applied electric field. The most direct evidence for moving charge-density waves comes from NMR experiments. Transitions to the CDW state have profound influence on the NMR spectrum, mainly due to the inhomogeneous broadening caused by the charge-density wave modulation through quad-

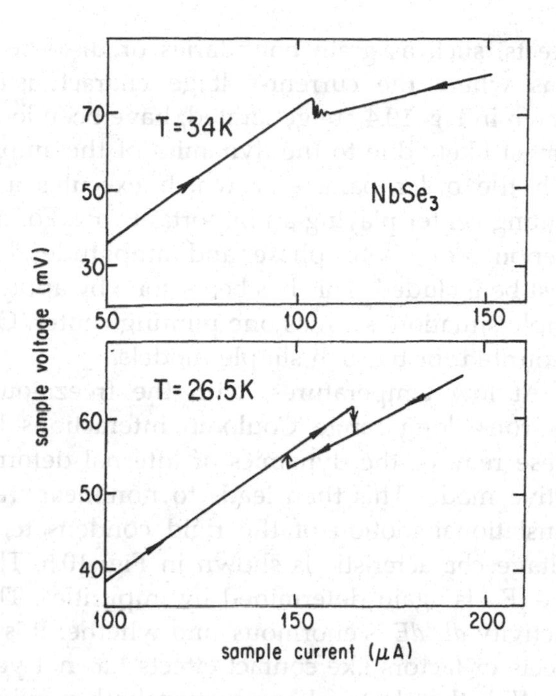

Figure 10.4. Current-voltage characteristics observed in $NbSe_3$ at two different temperatures (Zettl and Grüner, 1982). The behavior is due to extended defects, such as gain boundaries in the specimen.

rupole effects. The NMR line shape in the presence of uniform CDW velocity has been worked out in detail as a function of the velocity v. The solution leads to a motional narrowing and the appearance of a central line at the unperturbed-perturbed Larmor frequency. Experiments on $NbSe_3$ (Ross et al., 1986) and $Rb_{0.3}MoO_3$ (Nomura et al., 1986; Ségranson et al., 1986) clearly established the motional narrowing associated with moving charge-density waves. A detailed analysis and the evaluation of CDW velocity require calculation of NMR line shapes. When this is done, the current is found to be caused by the translational motion of the condensate.

In several cases, the onset of nonlinear conduction is accompanied by switching from the pinned to the current carrying state, and the behavior observed first in $NbSe_3$ (Zettl and Grüner, 1982) is shown in Fig. 10.4. Experiments employing contacts which are moved to different positions along the conducting directions have established that this behavior is due to extended defects, such as grain boundaries or dislocations; and the positions where the current-voltage characteristics, such as those shown in Fig. 10.4 are generated, have been located. The behavior is most likely due to the dynamics of the amplitude fluctuations, with the order parameter which executes a phase slip at the pinning center playing an important role. For a description of the phenomenon, both phase and amplitude degrees of freedom must be included. This has been done by applying Eq. (10.29) to a simple situation, such as one pinning center. Observations can be accounted for by such simple models.

At low temperatures, with the freeze-out of uncondensed electrons, long range Coulomb interactions become important. These remove the dynamics of internal deformations of the collective mode. This then leads to nonlinear transport due to the translational motion of the rigid condensate; a typical current-voltage characteristic is shown in Fig. 10.5. The sharp threshold field E_T is again determined by impurities. The differential conductivity dI/dE is enormous, and whether it is limited by intrinsic effects or factors like contact effects has not yet been established.

With the electrical conduction both nonlinear and frequency dependent, a broad variety of experiments can be performed in the presence of both *dc* and *ac* applied fields

$$E = E_{dc} + E_{ac} \cos \omega t. \tag{10.31}$$

Both the resulting *dc* and *ac* current can be observed. Experiments with small *ac* electric fields in general are straightforward to account for, in this limit the perturbative treatment of the nonlinear dynamics in terms of the *ac* drive is more appropriate. Such experiments, which include rectification and harmonic mixing can be described using the (experimentally established) frequency dependent and *dc* nonlinear characteristics (Grüner, 1988). The responses to large amplitude *ac* drives, such as depinning by *ac* fields are more complicated. Here qualitative arguments (Grüner et al., 1981) have been used to account for the experimental findings.

The observations on spin density waves are similar to those made on charge density waves (Grüner, 1993, and references cited therein). In materials like $(TMTSF)_2PF_6$, a sharp threshold field E_T is observed (Tomic et al., 1989), the magnitude of which is determined by the impurity concentration. Switching and hys-

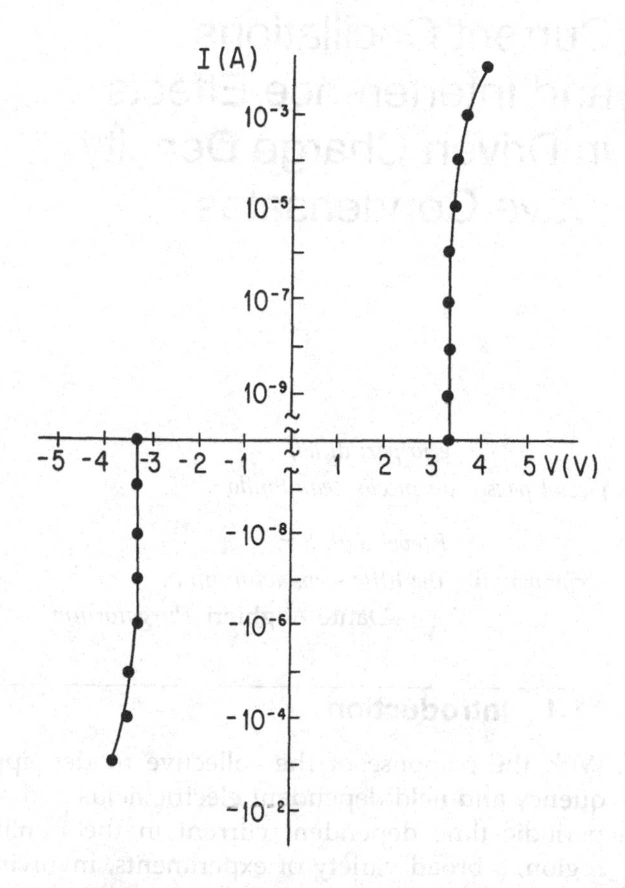

Figure 10.5. Current-voltage characteristics observed in $K_{0.3}MoO_3$ at low temperatures where screening by the uncondensed electrons is absent. The full line is a guide to the eye.

teresis effects, similar to those shown in Fig. 10.4, have also been found (Grüner, 1994 and references cited therein), presumably also due to pinning by extended defects such as grain boundaries. At low tempertures a different type of current-voltage characteristic has been observed (Mihály et al., 1992; Grüner, 1994). The results are suggestive of a tunneling process, which is different from single particle Zener tunneling; and which occurs at a significantly lower characteristic field than the fields which result from Zener tunneling across the gap.

Current Oscillations and Interference Effects in Driven Charge Density Wave Condensates

e io pari di lei,
picciol passo con picciol sequitando...

I level with her,
accompanying the little steps with mine...
—Dante Alighieri *Purgatorium*

11

11.1 Introduction

With the response of the collective modes applied to both frequency and field dependent electric fields and also because of the periodic time dependent current in the nonlinear conductivity region, a broad variety of experiments, involving the application of *dc* and *ac* drives can be performed. These experiments can be classified as follows:

1. **Frequency and electric field dependent conductivity** where the time average current $\langle j(t) \rangle$ is measured in the presence of *dc* or small amplitude *ac* drives. The *dc* conductivity is defined as $\sigma_{dc} = \langle j(t) \rangle / E_{dc}$ where E_{dc} is the applied electric field; and the small amplitude *ac* response is characterized by $\sigma_{ac} = \mathrm{Re}\{\sigma(\omega)\} + i\,\mathrm{Im}\{\sigma(\omega)\}$, where $\mathrm{Re}\{\sigma(\omega)\}$ and $\mathrm{Im}\{\sigma(\omega)\}$ are the real and imaginary components of the conductivity. Here it is assumed that linear response theory is appropriate, and the response to a sinusoidal *ac* drive $E_{ac} = E_0 \sin \omega t$ is also sinusoidal, with no higher harmonics. Because the *dc* response is nonlinear, it is obvious that this is not valid for arbitrary *ac* fields, and nonlinear *ac* effects may occur for finite fields. These experiments serve as

tests of the various models leading to predictions concerning the *dc* and *ac* response of charge density waves. The *ac* response can also be used to evaluate the fundamental parameters of the problem, such as the effective mass m^*, the pinning frequency ω_0, and damping constant $1/\tau$. Frequency dependent conductivity studies have been performed over a rather broad frequency range, from audio to sub-millimeter wave frequencies. By now the ω dependent response of the pinned mode is fully explored in many materials. Similarly, the *dc* conductivity has been explored in detail and compared with various theories. Experiments performed at finite fields and frequencies have also been conducted. The nonlinear *ac* response has been examined in detail in various materials. Such studies are supplemented by the measurements of rectification and harmonic mixing performed over a broad range of frequencies and applied *dc* and *ac* fields. These studies will not be summarized here; it should be mentioned, however, that (perhaps not surprisingly) models which account for the *dc* and small amplitude *ac* response are successful in describing the response in finite fields and frequencies, as well as in the presence of simultaneous *ac* and *dc* excitations.

2. Spectral response in the current carrying state. In the nonlinear conductivity region, the spectral features of the current include large amplitude broad band noise and current oscillations (often called narrow band noise, NBN), with a fundamental given by Eq. (10.20) and several harmonics $f_n = nf_0$ with slowly decaying intensity. The current oscillations have a finite spectral width and also display temporal fluctuations which can be studied by examining the time dependence of the Fourier transformed current. The origin of the current oscillations has been studied by using imaginative lead configurations for current and voltage contacts and also by employing thermal gradients to break up the coherence which leads to current oscillations. The observation of current oscillations is suggestive of significant coherence throughout the specimens, and it is expected that the range of current-current correlations $\langle j(t,\vec{0}), j(t,\vec{r})\rangle$ is comparable to the dimensions of the samples. The finite widths associated with the current oscillations (when viewed by employing a spectrum analyzer which detects the Fourier transformed current) and broad band noise cannot be explained in terms of external noise representing the dynamics of the internal degrees of freedom within the condensate.

3. **Interference experiments** where both *dc* and *ac* fields are applied, and where either the *dc* of the *ac* response is measured. Many of the observations are similar to those made on Josephson junctions, but the relevant frequencies for CDW dynamics are in the radiofrequency rather than the microwave spectral range. Features of mode locking between the intrinsic current oscillations and the externally applied *ac* field are closely related to the features of the current oscillations; additional effects, such as synchronization and coherence enhancement by external drives, can also be investigated by the joint application of *dc* and *ac* drives. Also, in contrast to microwave signals, waveforms other than sinusoidal can be applied at radio frequencies; these can be used to examine phenomena such as transitions from the pinned to the current carrying state in detail.

This chapter focuses on current oscillations and interference phenomena in an attempt to give an overview of progress in the field, and to summarize the open experimental and theoretical questions.

11.2 Current Oscillations

11.2.1 General Features

The current-voltage characteristics, and cordial conductivity $\langle \sigma \rangle = \langle j \rangle / E$ of TaS_3 is displayed in Fig. 10.3. The term $\langle j \rangle$ refers to the time average *dc* current measured in the presence of an applied constant voltage, or to the applied *dc* current—depending on whether constant current or constant voltage configuration is used to measure the current voltage characteristics. (In principle, the current-voltage characteristics can be different depending on whether a constant current or constant voltage configuration is employed. This difference, however, has not been studied in detail.) In the nonlinear conductivity region, the current has, in addition to the *dc*, an oscillating component most conveniently studied by monitoring the Fourier transformed current. Whether an oscillating current or oscillating voltage is measured depends on the external conditions, both of which have

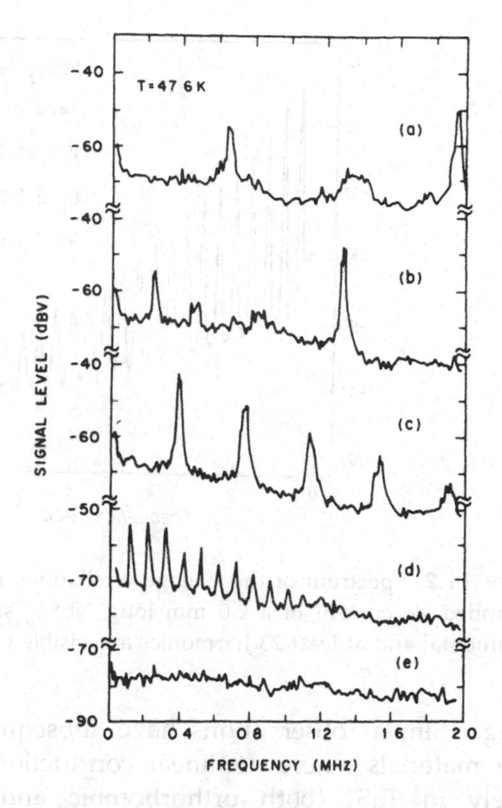

Figure 11.1. Fourier transform of the time dependent current in $NbSe_3$ for various applied currents. Narrow band "noise" results if the current exceeds the threshold value for nonlinear conduction. Currents and *dc* voltages are

(a) $I = 270\ \mu$A, V = 5.81 mV (b) $I = 219\ \mu$A, V = 5.05 mV

(c) $I = 154\ \mu$A, V = 4.07 mV (d) $I = 123\ \mu$A, V = 3.40 mV

(e) $I = V = .0$

The sample cross-sectional area $A \approx 136\ \mu m^2$ (Fleming and Grimes, 1979).

been observed. In the following, however, I refer to the phenomenon as current oscillation or "narrow band noise" (NBN); the latter is the traditional notation used. The first observation by Fleming and Grimes (1979) in $NbSe_3$ is shown in Fig. 11.1. A substantial broad band noise (BBN) accompanies the nonlinear conduction, and a superimposed narrow band "noise", with a fundamental and several harmonics is also observed. The spectrum moves to higher frequencies with increasing applied

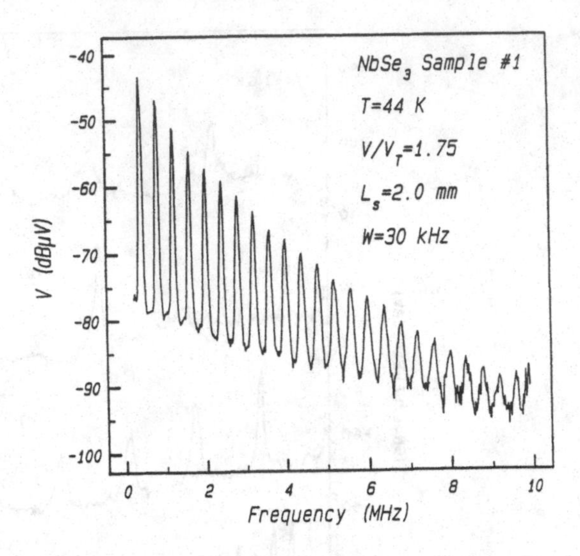

Figure 11.2. Spectrum of the voltage oscillations measured in response to an applied *dc* current of a 2.0 mm long $NbSe_3$ sample at 44 K. A single fundamental and at least 23 harmonics are visible (Thorne et al., 1987).

voltage. Similar observations have subsequently been made on other materials where nonlinear conduction has been observed, namely in TaS_3 (both orthorhombic and monoclinic phase), $(TaSe_4)_2I$, $(NbSe_4)_{3.3}I$ and $K_{0.3}MoO_3$ or $Rb_{0.3}MoO_3$. By now the phenomenon can be regarded as one of the fundamental observations of charge density wave dynamics.

The overall features of the observed current oscillations depend on crystal quality and in general on the coherence associated with the current carried by the collective mode. It has been established (Ong et al., 1984a, 1984b; Ong and Maki, 1985), that current or voltage contacts can introduce excess peaks in the Fourier transformed spectrum; similarly mechanical damage to the specimens also leads to complicated "noise" patterns. In carefully manipulated specimens the current oscillations have a rather high quality factor and also display many harmonics (Weger et al., 1980). Figure 11.2 shows the experimental results for $NbSe_3$. The coherence associated with the current oscillations can be characterized by the quality factor Q of the peaks in the Fourier transformed current. The quality factor $Q = \Delta_f/f_0$, where Δ_f refers to the width of the fundamental, was found to approach 10^5, in contrast to $Q \sim 4$ observed in Fig. 11.1.

As a general rule materials, where the electronic structure is more anisotropic, and consequently, the coherence perpendicular to the chain direction is less well defined, show broader current oscillation peaks; which are also less prominent for specimens with larger cross sections (Mozurkewich et al., 1983). These effects however, have not been investigated in detail. There is also a broadly defined correlation between the width of the narrow band noise and the amplitude of the broad band noise: in materials and in specimens of the same material with smaller broad band noise amplitude the Q of the Fourier transformed current peaks is larger. At one end of this spectrum, $NbSe_3$ has rather small broad band noise and high Q for the current oscillations in both CDW states; while $K_{0.3}MoO_3$ and $(TaSe_4)_2I$ both have large broad band noise at current oscillations with small quality factors. TaS_3 represents an intermediate case (Grüner, 1988). The quality of the crystals and of the voltage and current contacts can also be important (Beauchene et al., 1986) since macroscopic defects can lead to spurious velocity distributions in the specimens, while irregular contacts can lead to inhomogeneous current distributions.

Current oscillations have also been studied in the time domain by recording the time dependence of the current or voltage following a pulse excitation. Such experiments were performed in $NbSe_3$ (Fleming, 1983; Bardeen et al., 1982), and current oscillations have been subsequently observed in TaS_3 and $K_{0.3}MoO_3$ (Fleming et al., 1986). The observation of oscillations in the time domain differs from studies employing Fourier transforms in several aspects. First, for a finite Q of the oscillations a dephasing is expected, with a time dependent amplitude $\Delta j(f)$ given by

$$\Delta j(f) \sim A\left[1 - \exp\left(-\frac{t}{\tau}\right)\right] \tag{11.1}$$

with $\tau \sim Q^{-1}$. Such a gradual decrease of the oscillation amplitude has been observed in all materials, with a more dramatic decrease for oscillation with smaller Q values (Parilla and Zettl, 1985 and references cited therein; Fleming et al., 1985; Thorne et al., 1987c). Because of the fluctuating oscillation frequency, the time dependence of the oscillation amplitude most probably represents the dephasing of the various CDW domains, which are phase locked at the start of the applied pulse. Transient oscillations, induced by the sudden transition from the pinned to the current carrying state can also accompany the leading edge of the applied pulse. These transients have the fundamental oscillation frequency; the amplitude, however, is greatly enhanced over the steady state oscillation amplitude observed in the Fourier transform current under *dc* current or *dc* voltage drive conditions. Pulses of finite length of the order of few oscillation periods can also lead to mode locking, which will be discussed at length in Section 11.3. The observations can then be regarded as a form of interference and locking phenomena which are the consequences of competing periodicities; here the inverse frequency of the oscillations and the duration of the applied pulse.

11.2.2 Current-Frequency Relation

The frequency of the oscillations is approximately proportional to the current carried by the CDW (Monceau et al., 1980), and subsequently it has also been shown that the ratio of the current to the frequency varies with temperature, approximately as the number of condensed electron n_{CDW} (Bardeen et al., 1982). Fig. 11.3 shows the linear relation between I_{CDW} and f_0 in a wide frequency range for a pure $NbSe_3$ specimen (Bardeen et al., 1982); with the ratio I_{CDW}/f_0 as a function of T in the insert. Normalizing the CDW current for one chain, the above observations can be described as

$$\frac{j_{CDW}}{f_0} = ce\frac{n_{CDW}(T)}{n_{CDW}(T=0)} \tag{11.2}$$

where j_{CDW} refers to the current per chain carried by the condensate. While the precise value of the constant c in Eq. (11.2) is still debated it is between 1 and 2 for a broad range of materials. For $NbSe_3$, TaS_3 (both orthorhombic and monoclinic form) and for $(TaSe_3)_2I$, early experiments (Monceau et al., 1983) were suggestive of $I_{CDW}/f_0 = e$; indicating that the fundamental periodicity is related to a displacement of the collective mode by a half wavelength $\lambda/2$. There are, however, several ambiguities in the analysis. In $NbSe_3$ only part of the conduction electrons are condensed

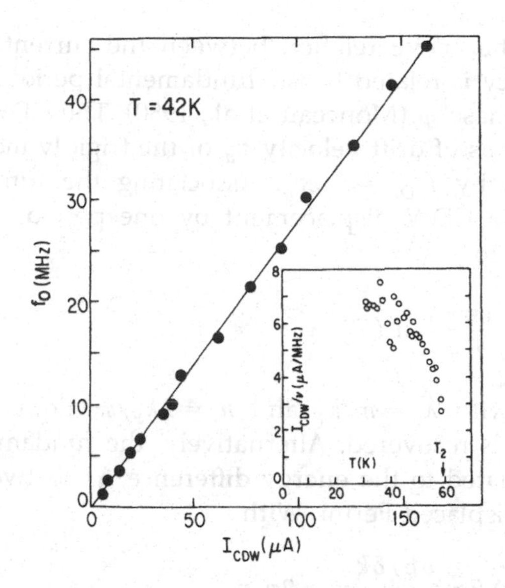

Figure 11.3. Relation between the CDW current and fundamental oscillation frequency in $NbSe_3$. The inset shows I_{CDW}/f_0 versus temperature (Bardeen et al., 1982).

into the collective mode, and the relation I_{CDW}/f_0 depends on the assumed number of condensed electrons. Similar problems arise in monoclinic TaS_3. The definition of the fundamental is complicated in $(TaSe_4)_2I$ where occasionally the second harmonic has a larger intensity than the fundamental. In orthorhombic TaS_3, the current-frequency relation in general is nonlinear due to the distribution of currents caused by the filamentary nature of the specimens. Using a mode locking technique to improve the coherence, the measured current-frequency relation in orthorhombic TaS_3 leads to (Brown and Grüner, 1985)

$$\frac{J_{CDW}}{f_0} = (2 \pm 0.1)e; \tag{11.3}$$

the above value was subsequently confirmed in an independent study which also employs mode locking (Latyshev et al., 1986). Careful NMR experiments on $K_{0.3}MoO_3$ specimens (Segransan et al., 1986) utilizing motional narrowing effects also lead to a relation $I_{CDW}/f_0 = (1.9 \pm 0.1)e$ in close agreement with the findings on orthorhombic TaS_3.

The above relation between the current and oscillation frequency is related to the fundamental periodicity associated with the phase ϕ (Monceau et al., 1980). The CDW current, described in terms of drift velocity v_d of the (rigidly moving) condensate is given by $j_{CDW} = n_c e v_d$. Associating the fundamental frequency with a CDW displacement by one period, $f_0 = v_d/\lambda_0$ this then leads to

$$\frac{j_{CDW}}{f_0} = n_c(T) e \lambda_0 \tag{11.4}$$

and with $\lambda_0 = \pi/k_F$ and $n_c = 2k_F/\pi$, Eq. (11.2) with a constant $c = 2$ is recovered. Alternatively, the fundamental frequency can be related to the energy difference ΔE between the two sides of the displaced Fermi. With

$$\Delta E \sim \frac{\delta E/\delta k}{\delta k/\delta v} v_d = 2 v_F v_d \tag{11.5}$$

$\Delta E = hf_0$ and $j_{CDW} = n_c e v_d$ also leads to Eq. (11.2) with $c = 2$. The strictly linear relation between j_{CDW} and f_0 as shown in Fig. 11.3, also suggests that the CDW velocity is constant throughout the specimen. Indeed j_{CDW} is proportional to f_0 over a broad range of currents and frequencies only in high quality materials. In materials where other, independent evidence suggests that disorder plays an important role, deviations from the linear relation between j_{CDW} and f_0 are observed (likely related to finite velocity-velocity correlation lengths) (Brown and Grüner, 1985).

The amplitude of the oscillating current increases with increasing collective mode current and appears to saturate in the high electric field limit (Weger et al., 1982; Mozurkewich and Grüner, 1983). This has been studied in detail in $NbSe_3$, and the amplitude of the oscillating current is displayed in Fig. 11.4 as a function of oscillation frequency. In experiments where the oscillation amplitude was measured up to $f_0 = 700$ MHz (Weger et al., 1982; Thorne et al., 1986a), the oscillation amplitude was found to saturate in the large velocity limit.

11.2.3 Size Effects and Fluctuation Phenomena

The current oscillations are suggestive of translational motion of the condensate in a periodic potential. Considerable effort has been made to gain information on the origin of the pinning by

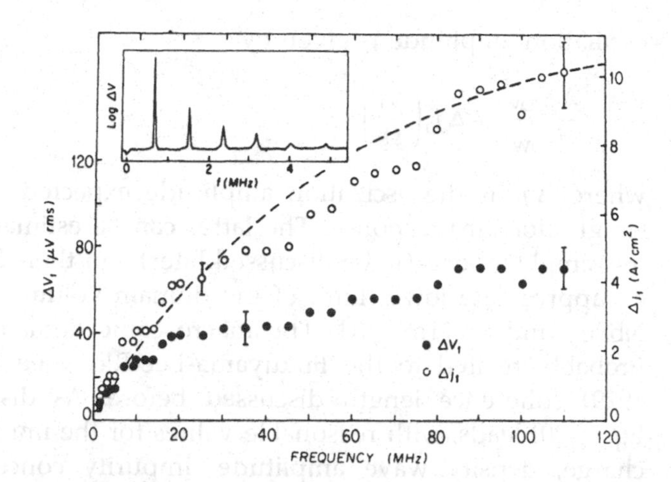

Figure 11.4. Amplitude of the fundamental oscillation in $NbSe_3$ at 42.5 K, vs. oscillation frequency. The amplitude is plotted both as a voltage ΔV_1 (solid symbols, left-hand scale) and as an equivalent current density Δj_1 (open symbols, right-hand scale). The inset shows a typical spectrum. The dotted line follows from the equation of motion Eq. (10.3) (Mozurkewich and Grüner, 1983).

measuring the oscillation amplitude as a function of sample volume; as well as under nonequilibrium conditions such as those provided by temperature gradients applied to the specimens. Although all experiments indicate that the current oscillations disappear in the thermodynamic limit (Mozurkewich and Grüner, 1983) several questions remain, mainly centered around the detailed mechanisms of pinning and the mechanism of oscillating current generation.

Experiments performed both in $NbSe_3$ (Mozurkewich and Grüner, 1983) and in $(TaSe_4)_2I$ (Mozurkewich et al., 1984) specimens with different lengths and cross sectional areas are in agreement with a volume dependence of the relative oscillation amplitude

$$\frac{\Delta j_{CDW}}{j_{CDW}} = \text{const.}\,\Omega^{-1/2} \qquad (11.6)$$

where Ω is the volume of the specimen. The results can be understood by postulating that the specimens are composed of domains of size L^3 where the current is coherent, but the currents in different domains oscillate with random phase. In this case, the oscillation amplitude is given by

$$\frac{\Delta j_{\mathrm{CDW}}}{j_{\mathrm{CDW}}} = \Delta j_0 \left(\frac{L_3}{\Omega} \right)^{1/2} \tag{11.7}$$

where Δj_0 is the oscillation amplitude expected for a coherent (single domain) response. The latter can be estimated by assuming simple models (to be discussed later) and then $L^3 \sim 0.2 \mu m^3$ as an appropriate lower limit of the domain volume is obtained in $NbSe_3$ and in $(TaSe_4)_2I$. The macroscopic domain size is most probably related to the Fukuyama-Lee-Rice (see Lee and Rice, 1979) coherence length discussed before. As discussed earlier Eq. (8.20) leads, with reasonable values for the impurity potential charge, density wave amplitude, impurity concentration, and Fermi velocity, to $L_0 = 100\mu$ along the chain direction. For an anisotropic band structure, the coherence length is also anisotropic (Lee and Rice 1979), and in the directions perpendicular to the chains the phase-phase correlation length $L_\perp$ is given by

$$L_\perp = \frac{v_{F\perp}}{v_{F\parallel}} L_\parallel . \tag{11.8}$$

A typical anisotropy of $v_{F\parallel}/v_{F\perp} \sim 30$ leads to $L_\perp \sim 0.3 \mu m$ and to a domain size $V = L_0 L_\perp^2$ of about $1 \mu m^3$, in broad agreement with the experimental results. A different set of experiments (Ong and Maki, 1985; Ong et al., 1984), performed on $NbSe_3$ however, indicates that for long specimens, the oscillation amplitude is independent of ℓ, the length of the specimens. This observation was used to argue that the oscillating current is generated at the contacts and not at the bulk (see also Jing and Ong, 1986). Subsequent experiments (Jing and Ong, 1986) where a length dependent oscillation amplitude was recovered, were also used to establish the range of phase coherence, and in $NbSe_3$ the experimental results are suggestive of $L_0 \sim 200 \mu$. The difference between the two sets of experiments may reflect the difference in contact configurations, impurity concentration, etc. The issue was further investigated by Brown and Mihály (1985) who employed a sliding, nonperturbative contact configuration to measure the noise amplitude in different sections of the specimens investigated. The observations are consistent with independent domains oscillating with the same frequency, but with random phase, as suggested by Eq. (11.6). In several specimens evidence was also found for sudden jumps in the oscillation amplitude when the contacts were moved along the specimens, indicating oscillating current generation by macroscopic defects such as grain boundaries within the specimens.

Several studies have been made to determine whether the oscillating current is generated in the bulk or at the contacts by applying a temperature difference between the two ends of the specimens. The argument for conducting such experiments is simple: with the two contacts at different local temperatures T_1 and T_2, they are expected to generate oscillations at different frequencies; each of which corresponds to the local temperature. Narrow band noise generated in the bulk, however, may be smeared by a temperature gradient; or alternately split into several frequencies each corresponding to a velocity coherent domain. The first experiments (Zettl and Grüner, 1984, 1985) on small $NbSe_3$ specimens did not reveal any splitting of the fundamental oscillation, which was also determined by the average temperature $(T_1 + T_2)/2$ of the specimen. The result is indicative of velocity coherence, comparable to the length $\ell \sim 1$ *mm* of the specimen. In subsequent experiments (Ong et al., 1985; Verma and Ong, 1984), however, a splitting was found and this was taken as evidence for current generation at the contacts. Several groups (Brown et al., 1985a; Lyding et al., 1986) later showed that by increasing the temperature gradient, further splittings can be induced, demonstrating that the various oscillation frequencies cannot be associated with contacts alone. In Fig. 11.5, the observations made by Lyding et al. (1986) are shown. A fundamental observed at $f = 1.1$ MHz for $\Delta T = 0$ is split into several peaks with increasing temperature gradient. The width of the peaks is approximately the same as that of the peak for $\Delta T = 0$, suggesting that the sample is broken up by the application of a temperature gradient into velocity coherent regions. The minimum velocity coherence estimated from the number of peaks and the sample length is of the order of 100 μm, in qualitative agreement with the conclusions of earlier studies.

The current oscillations, such as displayed in Figs. 11.1 and 11.2, have large temporal fluctuations both in position and amplitude. The time scale of the fluctuations is significantly larger than the inverse fundamental frequency, and is typically on the scale

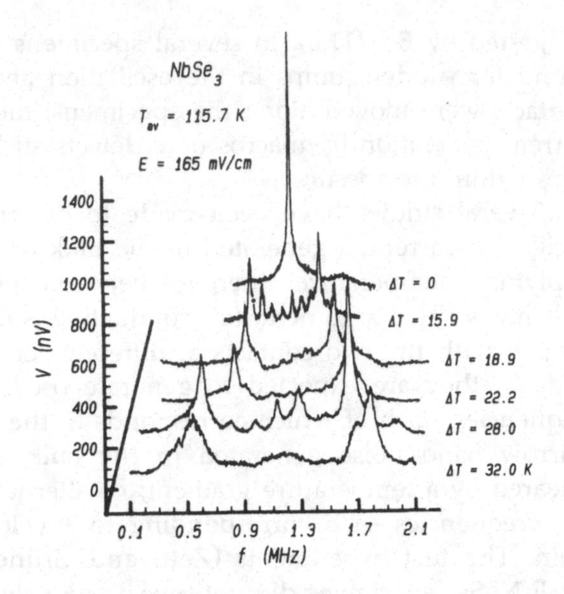

Figure 11.5. Splitting of the fundamental NBN peak for the upper Peierls transition of $NbSe_3$ as a function of applied thermal gradient. The gradient is opened around an average temperature of 115.7 K and the sample length is 0.89 mm (Lyding et al., 1986).

of microseconds. The phenomenon was first studied by Brown et al. (1985b) and investigated in detail by Link and Mozurkewich (1988) and by Bhattacharya et al. (1987).

The probability densities of the noise voltages follow a Gaussian distribution, suggesting that the current oscillations have many of the characteristics of Gaussian noise; and cannot be regarded as a coherent signal over a substantial period of time (Link and Mozurkewich, 1988). This is demonstrated in Fig. 11.6 where the probability distribution of the NBN voltages obtained for repeated scales of approximately 1 m is displayed. The full line is a Gaussian distribution with a width $\sigma = 3.3\ \mu V$. The distribution reflects the underlying probability density in the recorded oscillation amplitudes which are far enough removed from each other to be uncorrelated. This is appropriate for long sampling periods with strongly correlated oscillations expected for sampling with short time differences. A crossover from such coherent to Gaussian noise was also studied, and the typical time for the crossover is on the order of 10 μsec for $NbSe_3$. Similar experiments have not been performed in other materials,

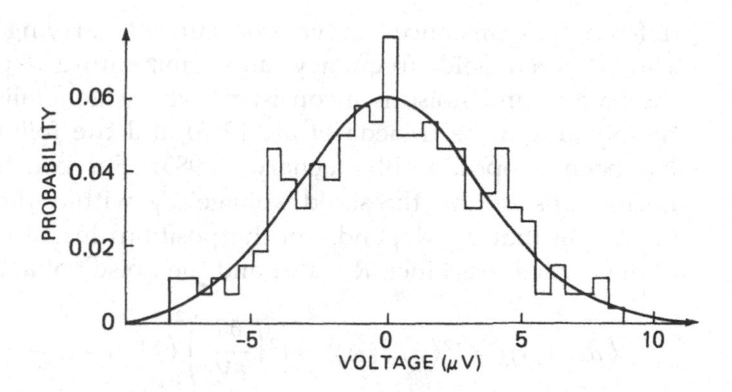

Figure 11.6. Histograms of voltages of narrow band noise in $NbSe_3$. The solid line represents the best fit Gaussian with $\sigma = 3.3\ \mu V$ (Link and Mozurkewich, 1988).

although temporal fluctuations of the oscillation amplitudes have been widely observed.

11.2.4 Broad Band Noise

The Fourier transformed spectra, such as displayed in Fig. 11.1, also have a significant broad band component, with an effective noise temperature corresponding to several thousand degrees of Kelvin (Richard et al., 1982; Zettl and Grüner, 1983). The amplitude of the noise is inversely proportional to the square root of the volume, $\Omega^{-1/2}$ (Richard et al., 1982), and the observation is suggestive of broad band noise generation in the bulk. In pure $NbSe_3$ specimens, however (Thorne et al., 1986b), the noise amplitude could clearly be related to the presence of macroscopic inhomogeneities (such as breaks in the specimens). This suggests that the broad band noise is generated at macroscopic boundaries such as grain boundary or breaks within the specimens. The issue here is similar to that raised in connection with the current oscillations; the most probable situation is that in general both macroscopic defects and impurities contribute to the broad noise since both lead to the disruption of the phase ϕ of the condensate. Clearly, detailed experiments on alloys with different impurity concentrations are required to clarify this point.

The spectral dependence can be described with a characteristic $\omega^{-\alpha}$ behavior with $\alpha < 1$ (Battacharya et al., 1985; Maeda et al., 1983, 1985) reminiscent of $1/f$ noise widely observed under different circumstances in various current carrying conditions. The observed field, frequency, and temperature dependence of the broad band noise is inconsistent with a generalized form of the Nyquist noise (Maeda et al., 1983) and the following model has been proposed (Bhattacharya, 1985). For constant current, fluctuations of the threshold voltage V_T within the specimens (assuming that E_T depends on the position) lead to fluctuations of the cordial resistance $R = V/I$ and the noise voltage is given by

$$\langle \delta V^2(\omega) \rangle = I^2 \langle \delta R^2(\omega) \rangle = I^2 \left(\frac{\delta T}{\delta V_T} \right)^2 \langle \delta V_T^2(\omega) \rangle \qquad (11.9)$$

It is also assumed that $\delta R/\delta V_T$ is proportional to $\delta V/\delta V_T$, i.e., R is a function of $V - V_T$ only, and that both quantities are independent of the frequency. (The latter is valid well below the pinning frequency ω_0.) Detailed measurements performed on TaS_3 indeed are in agreement with Eq. (11.9). The volume dependence of the noise indicates that it is an incoherent addition of fluctuations generated in coherent volumes, given by L^3. If the fluctuation of the effective pinning force δE_T is proportional to E_T, then

$$\langle \delta V_T^2(\omega) \rangle = \frac{E_T^2 L_0^3}{A} S(\omega) \qquad (11.10)$$

where $S(\omega)$ is the spectral weight function, $\int S(\omega)\, d\omega = 1$, and A is the cross section of the specimens. Eq. (11.10) is a consequence of the assumption that the number of independent entities generating the broad band noise is given by Ω/L^3. An analysis of the experimental results leads to $L_0^3 \sim 1\ \mu m^3$ in TaS_3, in broad agreement with coherent volumes estimated from the volume dependence of the current oscillations in $NbSe_3$ and in $(TaSe_4)_2I$ (Mozurkewich and Grüner, 1983; Mozurkewich et al., 1983).

11.3 Interference Phenomena

The nonlinear current-voltage characteristics and the observation of intrinsic current oscillations are suggestive of a variety of interference effects which may arise in the presence of combined *ac* and *dc* fields. In the case of Josephson junctions the phenomenon was used by Shapiro (1963) to demonstrate the existence of the *ac* Josephson effect and consequently many of the (formally similar) observations in driven charge density wave systems are also called "Shapiro phenomena."

Harmonic mode locking was first observed by Monceau et al. (1980). The observation was used to explore the relation between the oscillation frequency and time averaged current. The experiments were extended to explore the formal analog between CDW dynamics and observations made on Josephson junctions, including the characteristic Bessel function dependence of the interference features on the drive amplitudes (Zettl and Grüner, 1984). Subharmonic mode locking (Brown et al., 1984) has also been explored in detail together with incoherent effects, coherence enhancement (Sherwin and Zettl, 1985) and spatio-temporal effects (Bhattacharya et al., 1987). In contrast to Josephson junctions where the characteristic frequency is in the microwave spectral range, here *rf* frequencies are important, and consequently, waveforms different from a sinusoidal drive can be employed (Brown et al., 1986a, b).

11.3.1 Harmonic Mode Locking

Interference effects were first detected by Monceau and coworkers (1980) in $NbSe_3$ subjected to *dc* and *ac* applied fields. The differential resistance dV/dI measured in the nonlinear conductivity region with an *ac* field of varying frequency is shown in Fig. 11.7, for various current levels. Steps in the direct I-V curves correspond to peaks in the derivative; they occur when the fundamental frequency of the current oscillations (or its harmonics) coincides with the applied frequency. Although the spectrum of the interference peaks is complicated because of spurious effects (introduced most probably by inhomogeneous carrier injection at electrical contacts) a fundamental and several harmonics, such as those labeled by $4F_2$, $8F_2$, and $12F_2$ in the upper part of the figure, are evident. Spectra such as those shown in the figure were used to establish the linear current-frequency relations discussed earlier.

In carefully selected pure specimens, interference effects are also evident in direct I-V curves, such as that shown for $NbSe_3$ in Fig. 11.8 (Zettl and Grüner, 1984). The excitation applied to the sample had the form $V = V_{cd} + V_{ac}\cos(\omega t)$ with $\omega/2\pi = 100$ MHz. For $V_{ac} = 0$, a smooth, nonlinear I-V curve is observed with a well-defined threshold voltage V_T where the conductivity starts to be nonlinear. At higher values of V_{ac}, well-defined steps, which

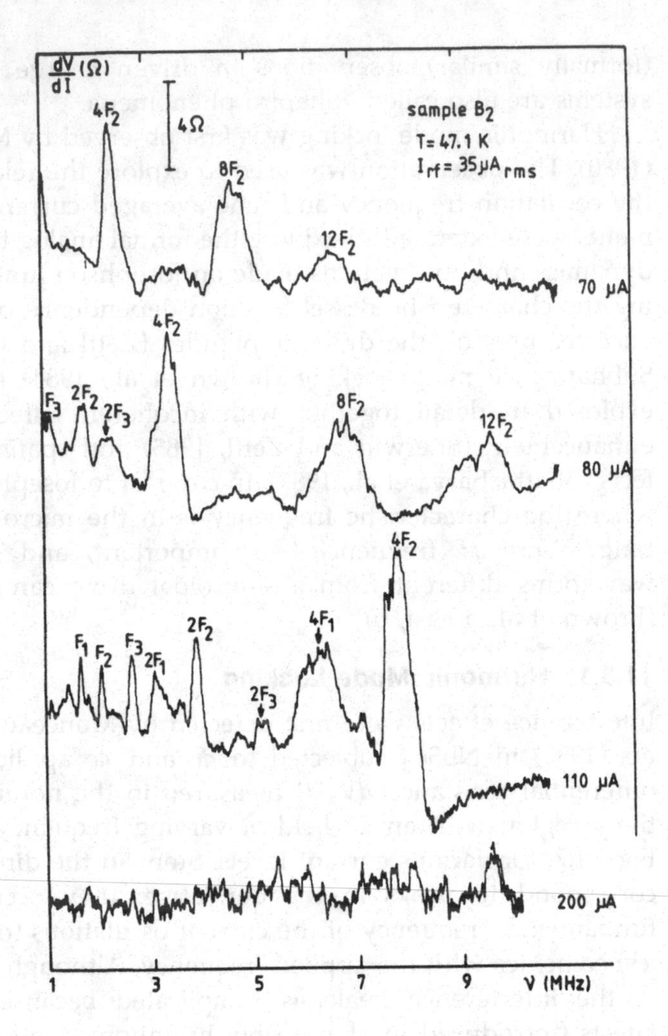

Figure 11.7. Differential resistance dV/dI in $NbSe_3$ observed by sweeping the frequency of the *rf* current, for a constant *dc* current. The various sets of interference peaks are labelled by F_1, F_2, F_3, and F_4 *dc* current (Monceau et al., 1980).

we shall index with an integer n, appear in the nonlinear region. The step height δV, as defined in the figure, in general first increases with increasing V_{ac} and then decreases. The position of the $n = 1$ step (identified in the figure) corresponds to a *dc* current I_{dc} which, in the absence of *ac* fields, yields an intrinsic oscillation of frequency $f_1 = 100$ MHz $= \omega/2\pi$. This has been established by measuring the oscillation frequency directly. We should also note the presence of harmonic steps corresponding to $n = 2$ (where $f = 200$ MHz) and smaller subharmonic steps corresponding to $n = \frac{1}{2}$ (where $f_1 = 50$ MHz). The steps are thus clearly an interference effect between the intrinsic current oscillation and the externally applied *rf* excitation.

Similar experiments have been performed on TaS_3 (Brown and Grüner, 1985) and also on $K_{0.3}MoO_3$ (Fleming and Schneemeyer, 1986). For both materials the current oscillations are in general smaller and have a significantly larger spectral width. These features are also evident in the interference patterns, shown for TaS_3 in Fig. 11.9.

The definition of the step height is straightforward, and is indicated in Fig. 11.8; its dependence on the *ac* amplitude

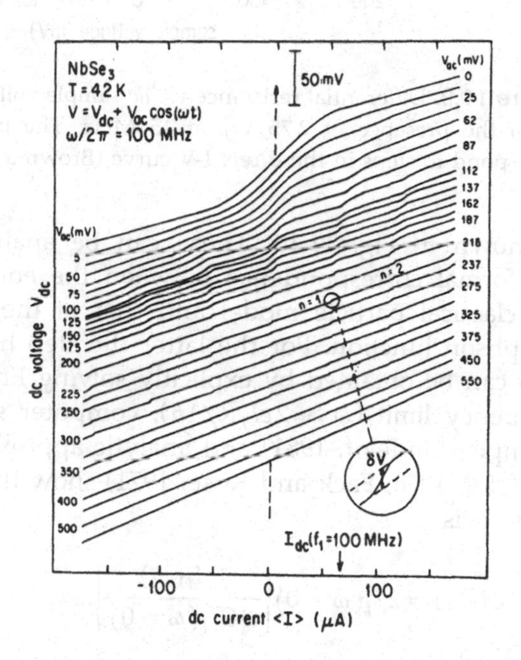

Figure 11.8. Shapiro steps in the *dc* I-V traces for $NbSe_3$ when *rf* field is applied at frequency $\omega/2\pi = 100$ MHz and amplitude V_{ac}. The step height δV is defined in the figure. No Shapiro steps are observed for $V_{ac} = 0$, while the maximum step height is at approximately $V_{ac} = 100$ mV. The arrow indicates the *dc* current which yields a fundamental noise frequency $f_1 =$ 100 MHz. The step index is n (Zettl and Grüner, 1984).

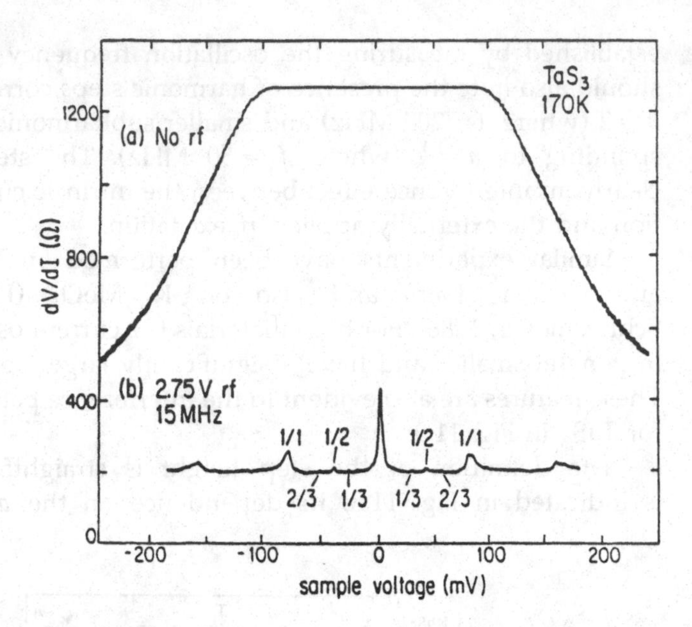

Figure 11.9. Differential resistance vs. *dc* sample voltage: (a) with no *rf* and, (b) in the presence of 2.75 *Vrf* at 15 MHz. The peaks in the derivative correspond to steps in the direct I-V curve (Brown and Grüner, 1985).

is shown in Fig. 11.10. These can be analyzed by employing the formal correspondence between the equation of motion of the classical particle model and that of the resistively shunted Josephson junction. For the latter the step height δI for the nth step can be obtained by explicitly solving Eq. (10.3). In the high frequency limit ($\omega \gg 2eI_J R/1\hbar$), computer simulations (see, for example, Lindelof, 1981) and analytic approximations (Clark and Lindelof, 1976; Fack and Kose, 1971) show that the height of the nth step is

$$\delta I(\omega) \approx 2I_J(\omega=0)\left|\frac{J_n I_I(\omega)}{\omega G I_J(\omega=0)}\right|. \tag{11.11}$$

where J_n is the Bessel function of order n. The critical current I_J depends on the applied *ac* current as

$$I_J(\omega)/I_J(\omega=0)|\,J_0(I_1(\omega)/\omega G I_J(\omega))|. \tag{11.12}$$

The corresponding equations for a CDW system are, by direct

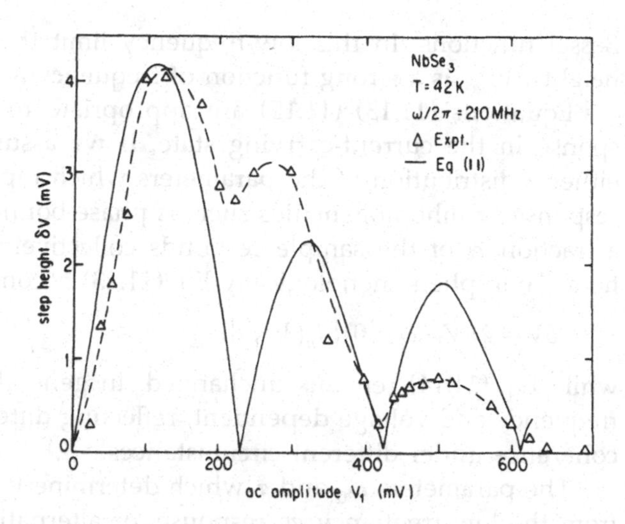

Figure 11.10. Step height δV versus *ac* amplitude V_{ac} for Shapiro steps in $NbSe_3$. The *rf* frequency is 210 MHz and the step index is $n = 1$. The solid line is the prediction of the classical model, Eq. (10.3), with parameters $\omega_0^2\tau/2\pi = 80$ MHz. $V_T = 24$ mV, $a = 0.17$ (Zettl and Grüner, 1984).

analogy to Eqs. (11.11) and (11.12)

$$\delta V = 2V_T(\omega = 0)|\, J_n(V_{\text{eff}})\,| \tag{11.13}$$

and

$$V_T/V_T(\omega = 0) = |\, J_0(V_{\text{eff}})\,|, \tag{11.14}$$

with V_{eff} given by

$$V_{\text{eff}} = \frac{V_{ac}\omega_0^2\tau}{\omega V_T(\omega = 0)}. \tag{11.15}$$

The above equations are appropriate in the high frequency limit for a strongly damped system where the capacitive term (for the Josephson junction case) or the inertial terms (for the CDW case) can be neglected. Equation (11.13) predicts that the maximum step height $\delta V_{\max}$ depends only on the maximum value of J_n and is thus independent of frequency ω. In the low-frequency limit, still neglecting the junction capacitance or CDW inertia; computer calculations (Lindelof, 1981) lead to solutions for the step heights in the I-V characteristics which closely resemble

Bessel functions. In this low-frequency limit the maximum step height δV_{max} is a strong function of frequency ω.

Equations (11.13)–(11.15) are appropriate for a coherent response in the current-carrying state. If we assume that (due to either a distribution of the parameters which represent the CDW response or inhomogeneities such as phase boundaries, etc.) only a fraction α of the sample responds collectively to the external field. Then phenomenologically Eq. (11.13) becomes

$$\delta V = 2\alpha V_T(\omega = 0)|\, J_n(V_{\text{eff}})\,|, \tag{11.16}$$

while Eq. (11.14) remains unchanged. In general, α can be both frequency and voltage dependent, reflecting different amounts of coherence under different circumstances.

The parameters ω_0 and τ which determine V_{eff} can be derived from the low frequency *ac* response, or alternatively, a fit to the data (such as shown in Fig. 11.10), can be used to calculate these parameters. The full line in Fig. 11.10, is Eq. (11.15) with $\omega_0^2\tau =$ 503 MHz and $\alpha = 0.17$. The former agrees well with $\omega_0\tau$ calculated directly from the frequency dependent response and the value of α indicates a highly coherent response (Zettl and Grüner, 1984).

The formalism also leads to a reduction of the threshold field which follows an oscillatory behavior (Lindelof, 1981), which was not observed in early experiments. Subsequent measurements on extremely well-defined specimens (Latyshev et al., 1987; Thorne et al., 1987b), however, clearly recovered the oscillatory behavior of V_T, as shown in Fig. 11.11.

Interference effects are also evident in the *ac* response measured in the nonlinear conductivity region. Fig. 11.12 shows the real and imaginary parts of the complex conductivity, Re$\{\sigma_{ac}(\omega)\}$ and Im$\{\sigma_{ac}(\omega)\}$, as a function of applied *dc* bias, measured at $\omega/2\pi = 3.2$ MHz. It is evident that neither the real nor the imaginary part of the frequency-dependent response is strongly affected by an applied *dc* voltage as long as $V_{dc} < V_T$. However, for $V_{dc} > V_T$, Re$\{\sigma_{ac}\}$ measured at 3.2 MHz, strongly increases and the *ac* dielectric constant ϵ, related to the *ac* conductivity by

$$\epsilon(\omega) = \frac{4\pi Im\{\sigma_{ac}(\omega)\}}{\omega} \tag{11.17}$$

strongly decreases for increasing V_{dc}. The dielectric constant ap-

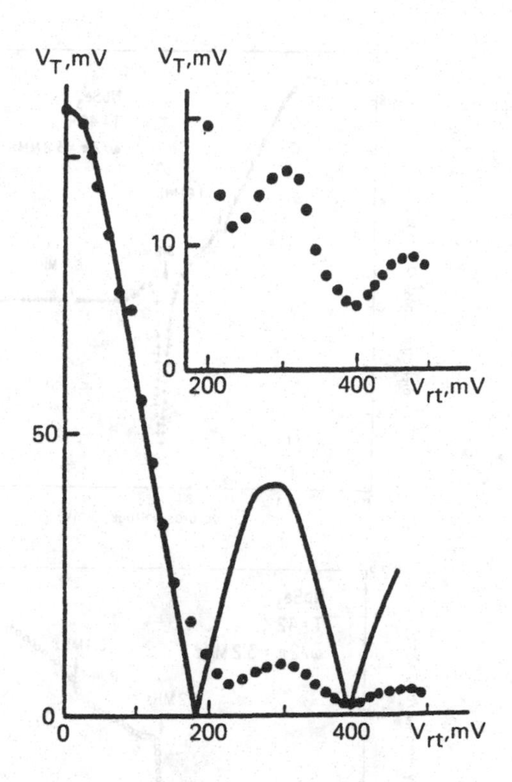

Figure 11.11. The *dc* threshold voltage versus *ac* amplitude for several applied *ac* frequencies. The period of the oscillations with *ac* amplitude is roughly proportional to the *ac* frequency. The solid lines are curves calculated by using Eq. (11.14) (Latyshev et al., 1987).

proaches zero for large *dc* drive showing that there is no appreciable out-of-phase component; Re$\{\sigma_{ac}(\omega)\}$, on the other hand, approaches the high-frequency limit obtained from the frequency dependence of the small amplitude *ac* response (observed for the pinned charge density wave).

In addition to the overall behavior described above, Figs. 11.12a and 11.12b also show that for $NbSe_3$ both Re$\{\sigma_{ac}(\omega)\}$ and $\epsilon(\omega)$ have sharp anomalies for well-defined values of V_{dc} in the nonlinear conductivity region. Specifically, Re σ_{ac} shows "steps" to higher conductivity values at $V_{dc} = 1.6$, 2.3, and 3.3 *mV*. At these same values of V_{dc}, ϵ shows well-defined inductive dips (Zettl and Grüner, 1984; Fleming et al., 1985). The *ac* response has not been calculated using the Josephson equation; but arguments

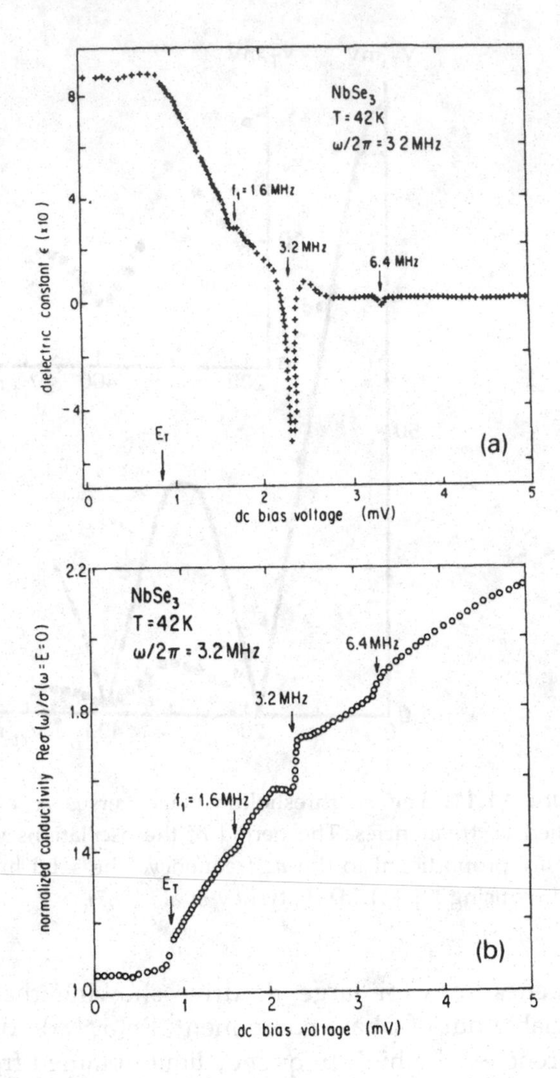

Figure 11.12. Real and imaginary part of the *ac* conductivity $\sigma(\omega)$ and dielectric constant $\epsilon(\omega)$ measured at $\omega/2\pi = 3.2$ MHz as a function of applied *dc* bias voltage. The threshold field is indicated by an arrow (Zettl and Grüner, 1984).

similar to those used in nonlinear circuit theory can be applied to account for the observations (Zettl and Grüner, 1984).

11.3.2 Subharmonic Mode Locking

Interference phenomena have also been observed whenever $pf_{\text{ext}} = qf_0$ for q an integer, but not equal to one (Brown et al., 1985). These are called the subharmonic Shapiro steps, and a rich array of such steps is shown in Fig. 11.13, again observed in $NbSe_3$. A few of the p/q values are identified in the figure; these were made by plotting I_{CDW} versus ω_{ext} and checking that the steps were p/q times that for the fundamental. In general, features with small p and q are more conspicuous and are both taller and wider. A complete mode locking would correspond to a plateau in dV/dI at the same level as the differential resistance below threshold. This can be understood as follows. The steps are regions of locking between the internal and applied frequencies when $pf_{\text{ext}} - qf_{\text{int}}$ is sufficiently small. If locking within such

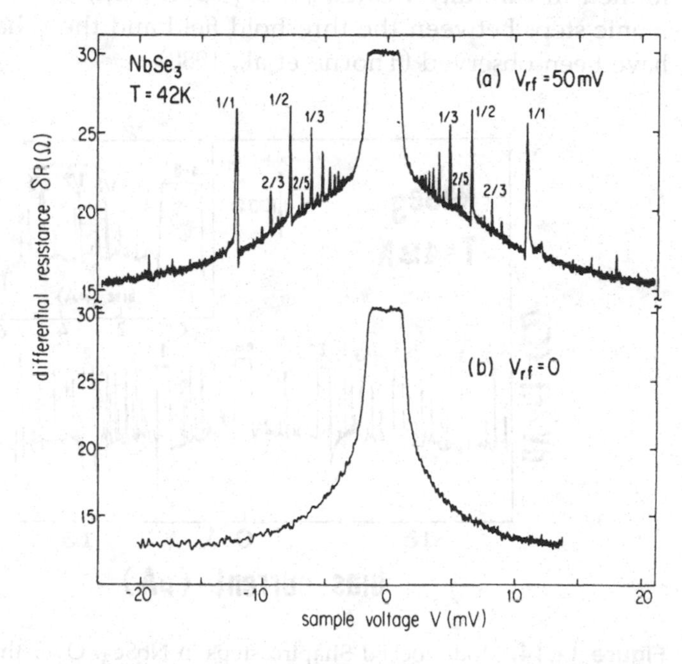

Figure 11.13. Differential resistance of $NbSe_3$ with and without an applied *rf* voltage V_{rf}. The numbers indicate the various subharmonic steps (Brown et al., 1984).

regions is complete, the CDW velocity becomes fixed by f_{ext} and does not respond to changes in the applied *dc* voltage. Hence dV/dI rises to the linear resistance R_0, which is solely due to the uncondensed electrons. This situation holds if the CDW velocity is coherent throughout the specimen. In reality the velocity coherence length must be finite, and f_{int} may vary spatially. If the variation of f_{int} is larger than the width of the region over which locking can occur, locking will be incomplete, and dV/dI will rise to level less than R_0. Observations under such circumstances are called interference "features." Therefore, the height of dV/dI is expected to correlate to the degree of synchronization across the sample. The completeness of synchronization depends on various factors such as sample dimensions, *ac* amplitude, and frequency; and in Fig. 11.14, subharmonic peaks several of which display complete mode locking, are shown (Hall and Zettl, 1984). While complete mode locking is observed only for a few subharmonic interference peaks in Fig. 11.14; in subsequent experiments performed in carefully treated $NbSe_3$ specimens, up to 150 subharmonic steps between the threshold field and the $\frac{1}{1}$ harmonic step have been observed (Thorne et al., 1988).

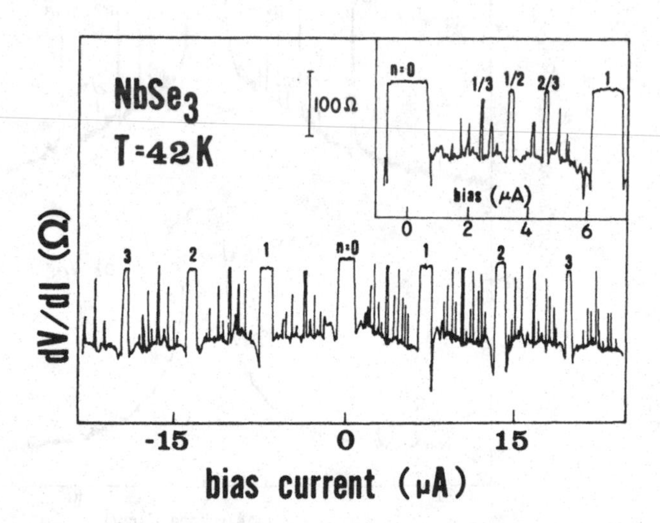

Figure 11.14. Mode locked Shapiro steps in $NbSe_3$. Over the mode locked region, dV/dI is independent of *dc* bias. The inset shows the subharmonic structure in detail, with corresponding p/s values (Hall and Zettl, 1984).

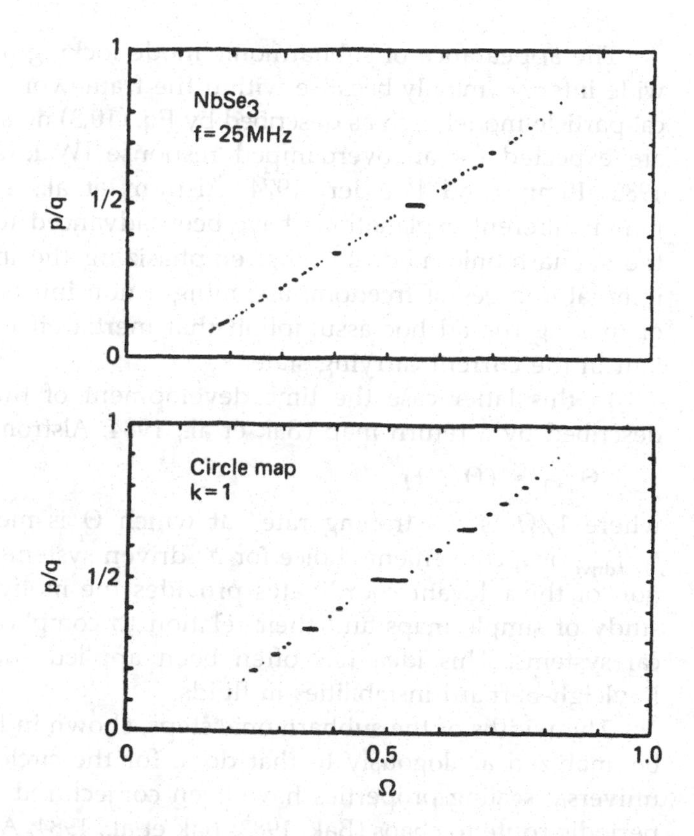

Figure 11.15. Widths of interference peaks for a $NbSe_3$ sample and for the circle map, with p/q marked on the vertical axis (Brown et al., 1986).

The detailed dependence of the subharmonic steps on the *ac* amplitude and frequency was also investigated, and characteristic experimental results are displayed in Fig. 11.15. Similarly to the Shapiro step corresponding to the fundamental (see Fig. 11.10), subharmonic Shapiro steps also display a characteristic Bessel function behavior, and in general have an amplitude which is reduced when compared to the fundamental Shapiro step. Although it is anticipated that an analysis in terms of Eq. (10.3) will provide an adequate explanation of the experimental results displayed in Fig. 11.15, such analysis has not been performed to date.

The appearance of subharmonic mode locking has generated wide interest, mainly because within the framework of the classical particle model such as described by Eq. (10.3) no subharmonics are expected for an overdamped response (Waldram and Wu, 1982; Renne and Poulder, 1974; Alstrom et al., 1984). Several rather different explanations have been advanced to account for the subharmonic mode locking, emphasizing the importance of internal degrees of freedom; assuming a nonsinusoidal potential or making the ad hoc assumption that inertia effects are important in the current carrying state.

In this latter case the time development of the phase δ is described by a return map (Bak et al., 1984; Alstrom et al., 1984)

$$\Theta_{n+1} = f(\Theta_n, \Omega) \tag{11.18}$$

where $1/\Omega$ is a "strobing rate" at which Θ is measured ($\Omega = f_1/f_{\text{drive}}$ is a convenient choice for *ac* driven systems). The reduction of the relevant coordinates provides the motivation for the study of simple maps and their relation to complicated dynamical systems. This idea has often been applied successfully for Rayleigh-Bernard instabilities in fluids.

The widths of the subharmonic steps, shown in Fig. 11.13, can be analyzed analogously to that done for the circle map, where universal scaling properties have been conjectured for the quasi-periodic route to chaos (Bak, 1983; Bak et al., 1984; Azbel and Bak, 1984). The circle map is a one-dimensional map with the following rule for the phase

$$\Theta_{n+1} = f(\Theta_n) + \Omega - (K/2\pi)\sin(2\pi\Theta_n), \tag{11.19}$$

and is considered a classic problem of competing periodicities, one coming from the phase space variable (Θ_1 is defined module 1) and the other from Ω. For any value of $K \leq 1$ the interactions of the map may converge to a limit cycle such that $\Theta_{n+q} = p + \Theta_n$ for a nonzero interval Ω (indicating a p/q subharmonic). When $K = 1$ every rational Ω will result in this kind of trajectory, and the plateaus form a staircase structure such as shown in Fig. 11.16. The staircase is said to be complete if the number of steps in an interval I of width greater than a discrimination level r obeys

$$N(r) = \frac{1 - S(r)}{r} \sim r^{-D},\ D < 1; \tag{11.20}$$

where $S(r)$ is the sum of the step widths that are greater than r.

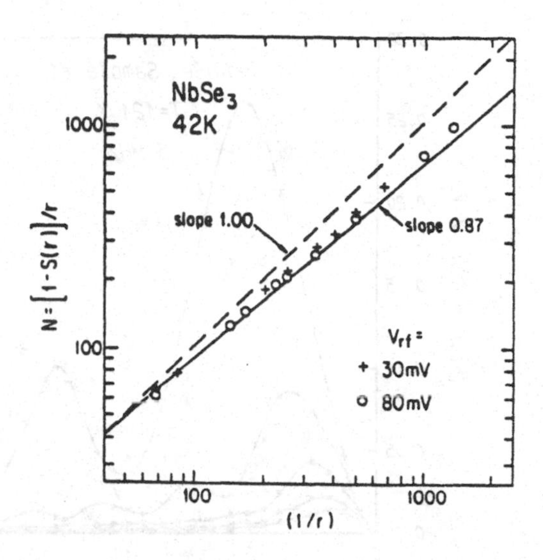

Figure 11.16. Experimental determination of the dimensionality D (see Eq. (11.20)).

The exponent D defines the (fractal) dimension of the staircase. For $K < 1$ the staircase is not complete and $D = 1$. At criticality ($K = 1$), $D = 0.87$ was calculated (Bak et al., 1984) and a similar value was found by analog simulations (Yeh et al., 1984), while for $K > 1$ the staircase structure breaks down and the motion is chaotic. A staircase, derived from differential resistance curves similar to that presented in Fig. 11.12, is also shown in Fig. 11.16. For both the circle map and $NbSe_3$ system, the smallest steps have been left out of the figure, but the two staircase structures are qualitatively similar, and in both cases the steps are in general larger for smaller q values.

The experimental results displayed in Fig. 11.12 have been used to evaluate the fractal dimension D (Brown et al., 1984). The construction $N(r)$ versus $1/r$ is shown in Fig. 11.17, leading to $D = 0.91 \pm 0.03$ in surprisingly good agreement with calculated values, and to those obtained from analog simulations. There are, however, several problems with such analysis. First, a complete staircase, $D < 1$ is expected only if the parameters of the system are measured at the phase boundary of the chaotic regime, i.e., for a well defined V_{ac} value. Experimental evaluations of D can also be limited by finite (instrumental or intrinsic) noise levels,

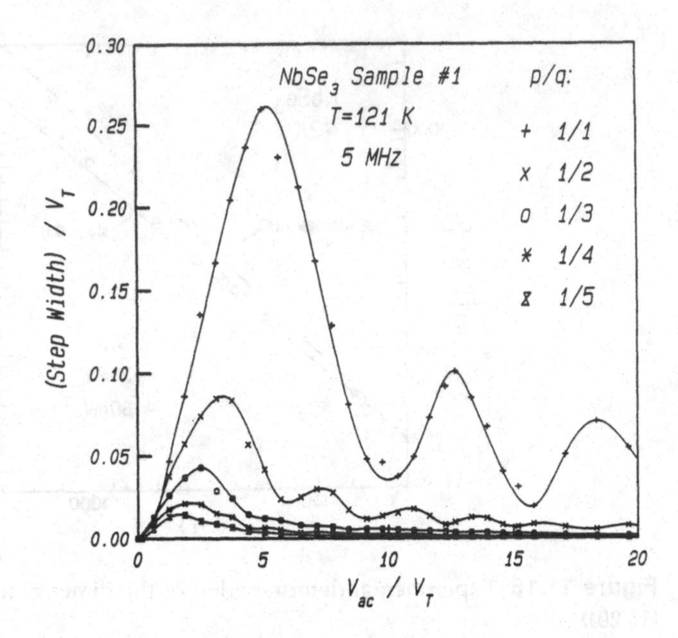

Figure 11.17. Widths of selected steps in the *dc* I-V characteristic versus peak *ac* amplitude for the applied *ac* frequency of 5 MHz. The period of the oscillations with *ac* amplitude varies as $1/q$ for the p/q step. The solid lines are guides to the eye (Thorne et al., 1987b).

smearing out the smaller steps. Also, computer studies of the sine circle map (Yeh et al., 1984), indicate that $D \sim 0.9$ can be obtained over a finite range of the parameter $K \leq 1$ if the smaller steps are not considered in the evaluation. This may occur for the CDW system as well. The good agreement between the fractal dimension, observed experimentally, and calculated on the basis of simple equations of motion, is most probably fortuitous, and cannot be regarded as evidence for a transition to chaos in the driven nonlinear system.

An alternative, conceptually simple explanation for the subharmonic steps has been advanced by Bardeen (see Thorne et al., 1986). The argument is based on a sinusoidal potential and employes arguments used originally by Shapiro (1963) to account for the steps in the Josephson junctions irradiated by microwave fields. The theorem that no subharmonics are expected in the overdamped response holds only for a sinusoidal potential which does not have higher harmonic Fourier components. Any other potential gives both harmonic and subharmonic locking, and consequently the observations are not surprising. The approach which is used to calculate the magnitude of the interference regions is the following (Thorne et al., 1986, 1987a). The current carried by the collective mode is given by equation

$$\phi(t) = \langle\phi\rangle_{L_0^3} \int_{L_0^3} \phi \, dr^3$$

and the time dependence of the average phase is, in the presence of applied *ac* and *dc* fields

$$\frac{d\Theta}{dt} = -\omega_d + A\cos(\omega_{\text{ext}}t); \tag{11.21}$$

where ω_d is the drift frequency $\omega_d = v_d/\lambda_0$ and A is proportional to the amplitude of the applied *ac* current. The time dependence of the phase then is

$$\Theta(t) = \omega_d t + \left(\frac{A}{\omega}\right)\sin(\omega_{\text{ext}}t) + \Theta_0. \tag{11.22}$$

This pinning potential can be Fourier expanded

$$V(\Theta) = \frac{a_0}{2} + \sum_{q=1}^{\infty} a_q(q\alpha\theta) \tag{11.23}$$

where $\alpha = 1$ for a periodicity given by the wavelength λ_0. After some algebra

$$V(\theta) = \frac{q_0}{2} + \sum_{q=1}^{\infty}\sum_{p=\infty}^{\infty} a_q J_p\left(q\alpha\frac{A}{\omega}\right) x \cos[(p\omega - q\omega_0)t + q\alpha\theta_0] \tag{11.24}$$

where J_p is the Bessel function of order p.

The conditions of phase locking depend on the equations of motion for the collective mode, but in general phase locking occurs if the time average pinning energy in the locked state is smaller than in the unlocked state. Without locking the time average pinning energy $\langle V(\theta)\rangle_t = a_0/2$, and for $p\omega_{\text{ext}} = q\omega_d$ an additional polarization energy

$$I\langle V(\theta_0)\rangle = \sum_q a_q I_q\left(q\frac{A}{\omega}\right)\cos(q\theta_0) \tag{11.25}$$

is obtained. The polarization energy is less than zero for a range of $\theta_0 - \theta_m < \theta_0 < \theta_m$ and θ_m determines the width of the step. $V(\theta)$ as given by Eq. (11.24) has been used to evaluate θ_m for different p and q values. As expected, both the harmonics and the subharmonics display a characteristic Bessel function behavior in broad agreement with the experimental results. The period of the oscillations with *ac* amplitude A varies linearly with frequency and inversely with q also in agreement with the observations as indicated in Fig. 11.17. In general, the agreement between experiment and theory is satisfactory for a broad range of applied *ac* frequencies and amplitudes (Thorne et al., 1987b). A similar agreement would, however, be obtained by any potential which has the same periodicity as that of Eq. (5.14) and has smooth minima separated by cusps.

A rather different concept has been developed by Tua and Ruvalds (1985), by Littlewood (1986), and by Matsukawa (1987) by extending the models which incorporate the internal degrees of freedom (discussed in the previous section), to account for the interference phenomena observed. Both conclude that the internal degrees of freedom lead to subharmonic mode locking even under circumstances where single degree of freedom dynamics lead only to locking corresponding to harmonic frequencies. The classical dynamics of coupled domains has been used by Tua and Zawadowski (1984) to account for the current-voltage characteristics and for the finite size effects associated with the current oscillations. The extension of the model to combined *dc* and *ac* drives (Tua and Ruvalds, 1985) leads to subharmonic steps, and to an apparent complete Devil's staircase. The overall amplitudes of the subharmonics agree well with the experiments. The calculations also lead to a subharmonic structure which depends only slightly on the drive conditions, with no critical values dividing complete and incomplete staircase behavior; in striking agreement with the experimental observations. The amplitude of the steps was found to decrease with the increasing number of domains, and the steps disappear in the thermodynamic limit. This feature of the results is in clear contrast with the hydrodynamic treatment of Sneddon et al. (1982) which predicts that (while the current oscillations are a finite size effect) interference peaks are recovered in the thermodynamic limit. The origin of this disagreement is not fully understood.

The model where the internal degrees of freedom are incorporated following the Fukuyama-Lee-Rice model (Fukuyama and Lee, 1978; Lee and Rice, 1979) has also been used to study the response of charge density waves to a combination of *dc* and *ac* drives (Coppersmith and Littlewood, 1986). As discussed earlier, second order perturbation treatment in the hydrodynamic limit (Sneddon et al., 1982) leads to interference effects involving the fundamental and higher harmonics. Higher order perturbation theory leads to interference features (Coppersmith and Littlewood, 1985a, b), i.e., sharp peaks in the dI/dV curves; but not full mode locking with plateaus as displayed in Fig. 11.14. Complete mode locking is recovered by numerical simulations on finite size systems where the advance of the average phase $\langle \phi \rangle$ is monitored. The results of the calculations, which display advances $\Delta\phi/2\pi$ corresponding to various mode lockings p/q, are shown in Fig. 11.18. The detailed waveform near mode locking has also been calculated by Coppersmith and Littlewood. The overall tendency of the experimental results, which are suggestive of so-called interference "features" (i.e., not complete mode locking) at high frequencies, and complete mode locking with well defined plateaus in dV/dI curves at low frequencies (Brown and Grüner, unpublished); is well reproduced by calculations which take the dynamics of internal degrees of freedom explicitly into account. Interference curves, calculated on the basis of single degree of freedom classical dynamics (called GZC) model (see Eq. (4.2)) and on the basis of the Fukuyama-Lee-Rice model are shown in Fig. 11.19, together with experimental curves, shown as the full line. The absence of well defined "wings" in experimental and numerical simulations based on the FLR model are taken as evidence of the dynamics of the internal mode. Both harmonic and subharmonic mode locking are also recovered by perturbational analysis of the Fukuyama-Lee-Rice model in the presence of combined *ac* and *dc* electric fields (Matsukawa, 1987). The model leads to clear anomalies in the current-voltage characteristics. Whenever $nq = mp$, complete mode locking such as shown in Fig. 11.14, for example, has not been recovered. Although the reason for this is not clear at present, it may be related to the deficiencies of the model itself (Bardeen, 1988), or to the breakdown of the perturbation theory. The conclusion has been sharply

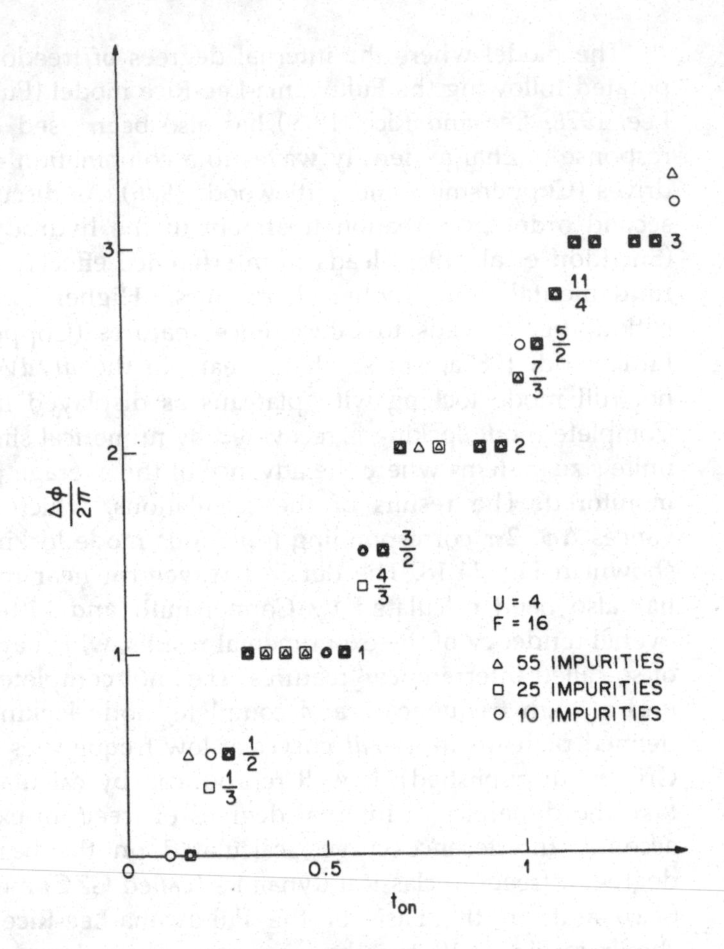

Figure 11.18. Number of wavelengths moved per pulse $\langle\Delta\phi\rangle 2\pi$ versus the pulse duration t_{on} with a fixed $E_{on} = 16$ for two systems with $U = 4$; one with 55 degrees of freedom and the other with 10. Locking is demonstrated because $\langle\Delta\phi\rangle 2\pi$ is always a rational fraction. The calculation is based on the Fukuyama-Lee-Rice model (Coppersmith and Littlewood 1986).

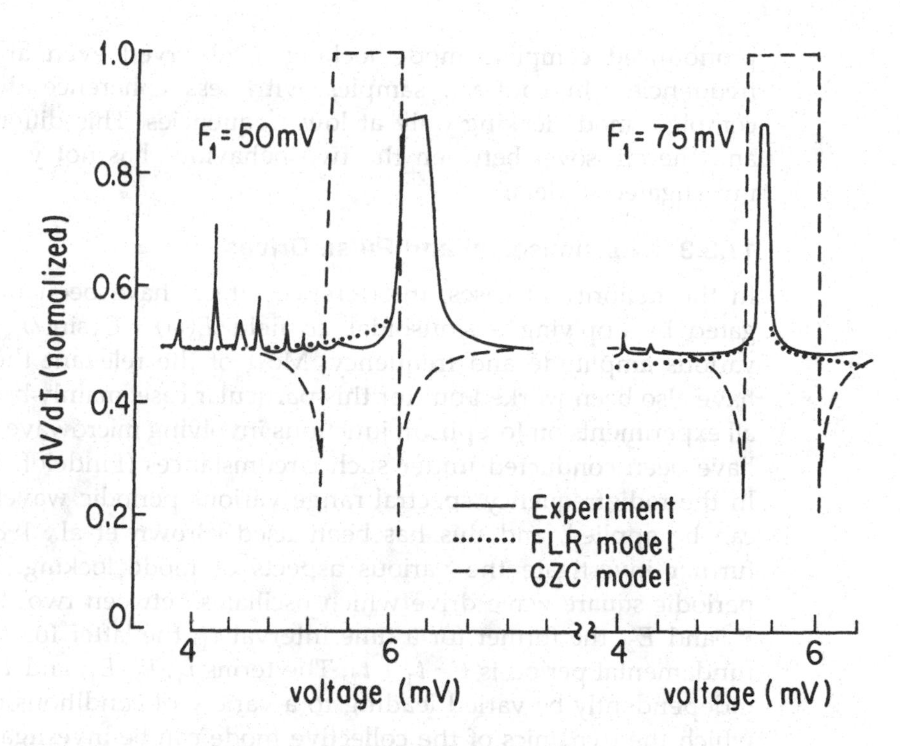

Figure 11.19. Differential resistance dV/dI plotted versus *dc* voltage V_0 near the first harmonic feature $Q.v \sim \omega_{AC}$ for $NbSe_3$ with *ac* frequency $\omega_{ac} = 25$ MHz and *ac* voltage amplitude $V_1 \sim 50$ and 75 mV. The sample's threshold voltage $V_T \sim 2$ mV. Also plotted are theoretical fits using the Fukuyama-Lee-Rice deformable model (dotted line) and the Grüner-Zawadowski-Chaikih one degree of freedom result (dashed line). The tops of the peaks are not calculated for FLR because the perturbation theory breaks down when the change in dV/dI is large (Coppersmith and Littlewood, 1986).

criticized recently by Thorne et al. (1987), who argued that complete mode locking can also be obtained at high frequencies, in apparent disagreement with the classical descriptions of charge density wave dynamics.

The disagreement between the various experimental results most probably reflects the difference in the degree of coherence in the materials investigated. As a rule, in small and pure specimens, where the phase-phase coherence length may exceed the dimensions of the samples; a highly coherent response, with pronounced complete mode locking is observed even at high frequencies. In contrast, samples with less coherence display complete mode locking only at low frequencies. This difference, and the crossover between the two behaviors, has not yet been investigated in detail.

11.3.3 Nonsinusoidal and Pulse Drives

In the majority of cases, interference effects have been investigated by applying a sinusoidal *ac* field $E(\omega) = E_0 \sin(\omega_{ext} t)$ of various amplitude and frequency. Most of the relevant theories have also been worked out for this particular case, mainly because all experiments on Josephson junctions involving microwave fields have been conducted under such circumstances (Lindelof, 1971). In the radiofrequency spectral range various periodic waveforms can be applied, and this has been used (Brown et al., 1986) to further investigate the various aspects of mode locking. For a periodic square wave drive which oscillates between two values, E_1 and E_2, the former for a time interval t_1, the latter for t_2, the fundamental period is $t = t_1 + t_2$. The terms t_1, t_2, E_1, and E_2 can independently be varied leading to a variety of conditions under which the dynamics of the collective mode can be investigated.

The difference between a sinusoidal drive and pulse drive, the latter with $t_1 = t_2$, is shown in Fig. 11.20. The sharp spikes in the upper part of the figure correspond to the fundamental interference peaks, with the $p/q = \frac{1}{2}$ subharmonics also evident on the figure. The amplitude of the interference peaks is approximately constant for the sinusoidal drive for small *dc* voltages, and starts to decrease when $|E_{dc} - E_{ac}| < |E_T|$; i.e., when the system is not driven back to the pinned configuration during the experiment. In contrast, the steps are suppressed near zero *dc* bias for the square wave drive, in the interval where the *ac* drive amplitude is large enough that the applied field drives the system from above threshold to above threshold (with opposite polarity) without allowing it to relax to the pinned configuration. As with the sinusoidal drive, the interference peaks are again suppressed at large E_{dc}, for which $|E_{dc} - E_{ac}| < E_T$. The above difference between the results for a sinusoidal and a square wave drive is clearly due to the fact that a finite time in the pinned state is required to observe substantial interference effects. While the overall features of Fig. 11.20 can be reproduced by computer

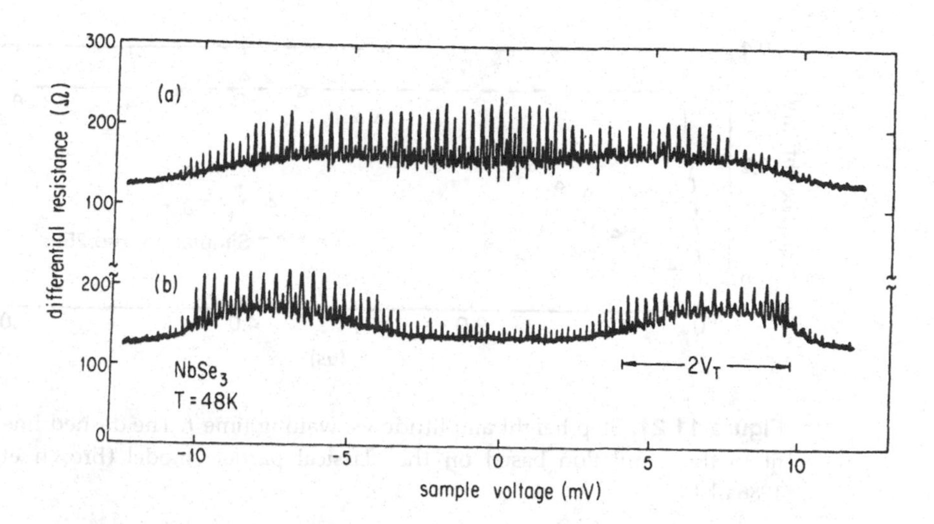

Figure 11.20. Differential resistance curves for (a) sinusoidal *ac* drive and (b) square wave drive for the same amplitude (Brown et al., 1986a, b).

simulations using only single degree of freedom dynamics (Brown et al., 1986a), experiments can also be conducted by varying t_q; the characteristic time spent below E_T, with $E_2 > E_T$ applied for time t_2 varied. Such experiments point to the importance of internal degrees of freedom in the dynamics of the collective mode. The measured step height is shown in Fig. 11.21 as a function of the time t_1 spent below threshold. As expected, the step height approaches zero as t_1 (called t_- in the figure) gets shorter and saturates in the long time ($t_1 > \infty$) limit. For short times the step height δV can be described by the expression

$$\delta V \sim V_T\left(\frac{t_1}{\tau_0}\right) \tag{11.26}$$

where τ_0 is the characteristic relaxation time of the system. For an overdamped classical particle $\tau_0 = \omega_0^2\tau$. In the long t_2 limit simple arguments (Brown et al., 1986a, b), based on the classical particle model lead to a step height

$$\delta V \sim V_T\left(\frac{t_1}{t_2}\right). \tag{11.27}$$

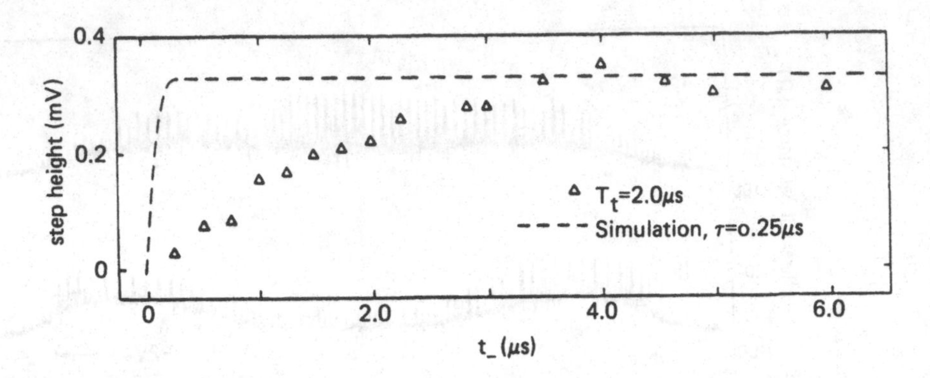

Figure 11.21. Step height amplitude vs. waiting time t. The dashed line is a fit to the simulation based on the classical particle model (Brown et al., 1986a, b).

Computer simulations, with $t_1 = 0.25\ \mu s$, obtained by fitting the observed ω dependent response to Eq. (11.27), lead to the dotted line in Fig. 11.21. The difference between the observed step height and that which follows from the classical particle model is significant, and is most probably a consequence of the dynamics of the internal degrees of freedom (Brown et al., 1986).

A different type of interference effect is called, in general, "pulse duration memory." The notion of pulse duration memory effect refers to the observation which is shown in Fig. 11.22 $K_{0.3}MoO_3$ (Fleming and Schneemeyer, 1986). Square wave voltage pulses which drive the system from the pinned to the current carrying state lead to transient current oscillations with frequencies given by Eq. (11.2). The amplitude of the oscillations gradually decreases with increasing time, and for $t \rightarrow \infty$ the magnitudes of the oscillations correspond to those which are measured under *dc* conditions by detecting the Fourier transformed current. This can be thought of as the consequence of gradual dephasing of the current oscillations which start with the same phase but which have slightly different frequencies depending on the local currents within the specimen. At the end of the pulse, there is a sharp upward cusp in the observed current, which occurs for a broad range of pulse durations t_0. The cusp suggests that the system adjusts itself to the pulse duration in such a way that its velocity is always rising as the pulse ends. Apparently, the system

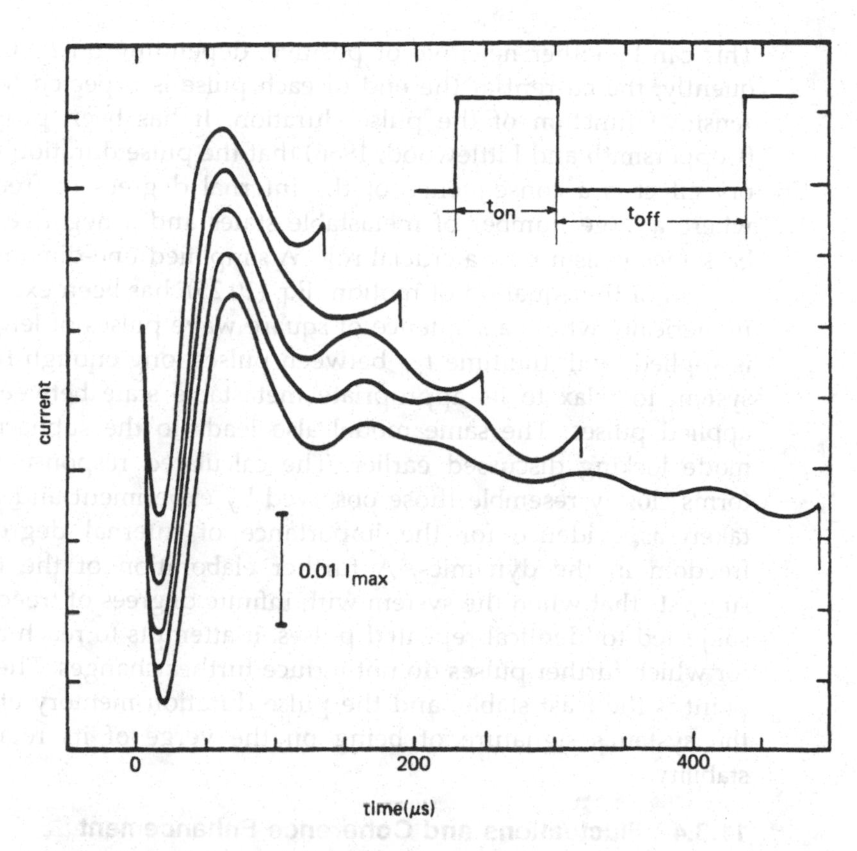

Figure 11.22. Current oscillations in response to a square-wave driving field of about $10E_T$ (inset) in $K_{0.3}MoO_3$ at 45 K. The data was obtained in a current driven configuration and has been inverted; however the current oscillations are also clearly observed in a voltage-driven configuration. These oscillations have also been reproduced by numerical simulations for a one-dimensional system with many degrees of freedom (Fleming and Schneemeyer, 1986).

"remembers" the length of preceding pulses, hence the name "pulse duration memory effect." The upward cusp, as shown in Fig. 11.22, is not observed for single pulses, although the current oscillations are evident.

Clearly a single particle description does not account for this finding as the current is determined by the derivative of the potential at the position of the particle at the end of each pulse.

This can be either negative or positive, depending on t_0. Consequently, the current at the end of each pulse is expected to be a sensitive function of the pulse duration. It has been proposed (Coppersmith and Littlewood, 1986) that the pulse duration memory effect is a consequence of the internal degrees of freedom, where a large number of metastable states and a negative feedback mechanism play a crucial role. A simplified one-dimensional version of the equation of motion, Eq. (10.25), has been examined numerically where a sequence of square wave pulses of length t_0 is applied; with the time $t_{\rm off}$ between pulses long enough for the system to relax to its appropriate metastable state between the applied pulses. The same model also leads to the subharmonic mode locking discussed earlier. The calculated response waveforms closely resemble those observed by experiment and this is taken as evidence for the importance of internal degrees of freedom in the dynamics. A further elaboration of the model suggests that when the system with infinite degrees of freedom is subjected to identical repeated pulses, it attempts to reach a state for which further pulses do not induce further changes. The fixed point is the least stable, and the pulse duration memory effect is the system's signature of being on the verge of its region of stability.

11.3.4 Fluctuations and Coherence Enhancement

Fluctuation effects associated with the current oscillations, discussed in Section 11.2, are suggestive of coupled domains of size L_0^3 which oscillate at approximately the same frequency $f_0 = j/2e$ with random phase. The Gaussian distribution of the voltages of the narrow band noise shown in Fig. 11.6, is suggestive of independent phase dynamics for the various domains. The mode-locking phenomena discussed earlier can be understood on the basis of the dynamics of a single degree of freedom system; however, the volume dependence of the current oscillation amplitudes suggest that the internal degrees of freedom, and the finite size effects are important (Mozurkewich and Grüner, 1983). Also, interference experiments performed for relatively small *ac* drives give mode locking smaller than that expected for a single degree of freedom dynamics, and this has been interpreted as partial mode locking (Zettl and Grüner, 1984). The reason for this is again most probably the absence of complete phase coherence throughout the specimens. It is then expected that phase synchronization by the applied *ac* field leads also to the reduction of incoherent phenomena, such as the broad band noise, or the narrow band noise fluctuations observed in the current carrying state. These have been investigated recently in detail (Bhattacharya et al., 1987; Sherwin and Zettl, 1985). The broad band noise spectrum is significantly reduced during mode locking, with a total noise power reduced to that comparable to that observed in the pinned CDW state (Sherwin and Zettl, 1985). This most probably is related to the freeze-out of the internal degrees of freedom, and may signal the enhancement of the dynamic coherence length which characterized the current-current correlations. Such an effect has been recovered by a perturbational analysis of the Fukuyama-Lee-Rice model in the presence of combined *dc* and *ac* drives (Matsukawa, 1987). The reduction is most probably the consequence of phase synchronization between the various domains. This reduces the dynamics to that of a single domain with no fluctuations and consequently no broad band noise. The noise reduction depends on the applied *ac* amplitude, as shown in Fig. 11.23; where the total measured noise power is plotted as a function of *ac* current in this current-driven experimental arrangement. The gradual reduction of the noise power is suggestive of a gradual phase homogenization, and a

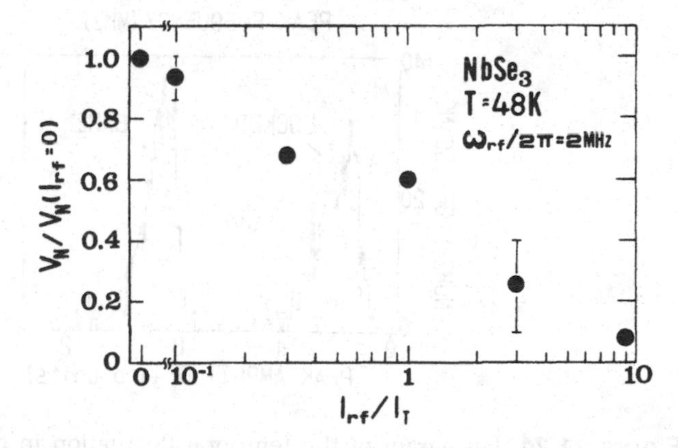

Figure 11.23. Total broad band noise amplitude vs. amplitude of an applied *rf* current in $NbSe_3$. Intense *rf* field suppresses the noise by homogenizing the CDW phase (Sherwin and Zettl, 1985).

complete phase homogenization is achieved for *ac* drive amplitudes which are typically about one order of magnitude larger than those which are required for complete mode-locking.

Fluctuations of the current oscillations are also influenced by the mode-locking, and both the amplitude and frequency fluctuations are reduced. The system synchronized at the $p/q = \frac{1}{2}$ subharmonic steps displays amplitude fluctuation reductions at short time scales, but may fluctuate between various mode locked states (characterized by different oscillation amplitudes) over a large time interval. No amplitude fluctuation reduction has been observed in a more detailed experiment (Bhattacharya et al., 1987) but the frequency distribution of the current oscillations was eliminated by mode-locking. The phenomenon is shown in Fig. 11.24. The upper part of the figure shows the fluctuations of the

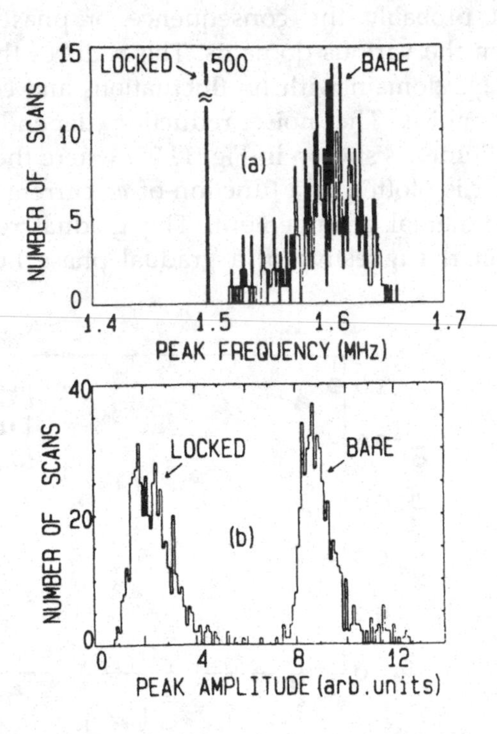

Figure 11.24. Histogram of the temporal fluctuation in frequency (a) and amplitude (b) of a "bare" NBN ($V_{ac} = 0$) and a locked NBN ($p/q = 1/2$ at $\omega = 3$ MHz). Results represent 500 scans of each case. In (b) the two histograms are shifted arbitrarily for clarity (Bhattacharya et al., 1987).

NBN in position and amplitude without the application of *ac* drive. When synchronization occurs, the NBN peak frequency does not fluctuate, but the amplitude fluctuation is unchanged. The absence of frequency distribution implies that at mode locking all temporal fluctuations of the oscillation frequency are quenched, i.e., the velocity degrees of freedom are eliminated.

Coherence enhancement and fluctuation suppression, such as observed in driven charge density wave systems, probably represent the most interesting consequences of nonlinear dynamics of many degree of freedom systems. Some aspects of this phenomenon, such as subharmonic mode-locking, under circumstances where a single degree of freedom system would not lead to subharmonic locking, have been explored. It remains to be seen whether or not the essential aspects of coherence can be accounted for.

11.4 Conclusions

The current oscillation phenomena which occur in driven charge density wave systems in the nonlinear conductivity region and associated interference effects in the presence of *dc* and *ac* drives are clear manifestations of a new type of collective transport phenomenon, carried by an electron-hole condensate. The linear relation between the time average current and oscillation frequency reflect the fundamental $2k_F$ periodicity associated with the electron hole condensate, and the mere existence of current oscillations in macroscopic specimens is suggestive of macroscopic length scales involved. The collective mode is characterized by an amplitude and phase; and the length scales are related to the phase-phase correlation length, determined by the impurity concentration and the parameters of the collective mode. The highly coherent response clearly demonstrates that they are comparable to the dimensions of the specimens investigated.

In spite of a broad variety of experiments performed in both the time and frequency domains and conducted by employing various combinations of *ac* and *dc* drive amplitudes, frequencies and waveforms, several unresolved questions remain. The linear relation between the time average currents and oscillation frequency is by now well confirmed in all materials which display

CDW transport phenomena; there is however, considerable uncertainty concerning the numerical factors involved. Experiments on TaS_3 suggest that simple arguments advanced in Section 11.2.2 are correct and the oscillation is related to the advancement of the phase by 2π, corresponding to the displacement of the collective mode by one wavelength. Careful experiments in other structurally simple model compounds, such as $(TaSe_4)_2I$ and $K_{0.3}MoO_3$ would be highly desirable. The relation, given by Eq. (11.2) holds also only at $T = 0$, and experiments on $NbSe_3$ indicate that the ratio j/f_0 is proportional to the number of condensed electrons. In contrast, in TaS_3 the ratio was found to be only weakly temperature dependent (Brown and Grüner, 1985). Further experiments, in particular the transition temperature near T_p, are needed to clarify whether the proposed relation

$$\frac{j}{f} = 2e\frac{n_{\mathrm{CDW}}(T)}{n_{\mathrm{CDW}}(T=0)} \tag{11.28}$$

is appropriate.

Considerable controversy exists concerning the spectral width of the current oscillations, and the related broad band noise. Broadly speaking, larger spectral widths are accompanied by larger broad band noise amplitudes suggesting that the two types of incoherent effects are strongly related; and that both are determined by the relative magnitudes of the phase-phase correlation length and dimensions of the specimens. One particular model (Bhattacharya et al., 1987) relates the broad band noise amplitude to the spatial and temporal fluctuations of the CDW current, through the relations $\langle \delta V^2 \rangle = \langle \delta I^2 \rangle R_N$; where R_N is the resistance due to the normal electrons. Fluctuations in CDW velocity also lead to fluctuations in the oscillation frequency explaining the above overall correlation. Studies of the broad band noise amplitude performed on TaS_3 and $NbSe_3$ as a function of the dimensions of the specimens indicate that it is a bulk effect and that noise generation is due to randomly positioned impurities. In contrast, studies on rather pure $NbSe_3$ specimens indicate that macroscopic defects, such as grain boundaries, are the main source of broad band noise. Macroscopic defects, and also random impurities, have also been suggested as the source of the current oscillations. The most probable explanation for the variety of findings is that in general, both impurities and extended boundaries (such as grain boundaries, contacts, etc.) contribute to the noise generation (both broad band and narrow band); but more detailed studies in alloys, or in irradiated specimens where the concentration of pinning centers is systematically varied, are called for to settle this issue.

Fluctuation effects, and clear evidence that the current oscillations disappear in the thermodynamic, infinite volume limit, demonstrate that descriptions of CDW dynamics within the framework of single degree of freedom models is not appropriate. Various experiments however, suggest that the dynamics of the collective mode are characterized by macroscopic length scales. The static phase-phase correlation length (called the Fukuyama-Lee length) L_0 is, for typical impurity concentrations and CDW parameters, on the order of 10–100 μm along the chain directions and one or two orders of magnitude smaller perpendicular to the chains. This leads to a typical "domain" size of 10^{-2} $\mu m - 1$ μm in broad agreement with estimates of the phase-phase correlation length on the basis of the length dependent threshold field (Monceau et al., 1986; Gill, 1982; Zettl and Grüner, 1984; Borodin et al., 1986), current oscillation amplitudes (Mozurkewich and Grüner, 1983), and broad band noise (Bhattacharya et al., 1985; Richard et al., 1982) implying that the static and dynamic aspects of the problem are characterized by the same length scale (except perhaps near threshold, see Fisher, 1985). As L_0 is expected to be inversely proportional to the impurity concentration, similar experiments on alloys would be of great importance.

Interference effects confirm many of the conclusions which have been reached on the basis of the current oscillation studies alone, but also emphasize the formal similarity between the Josephson phenomena and nonlinear CDW transport. While for small amplitude *ac* drives fluctuation effects, volume dependences for interference effects, and current oscillations go hand in hand; large amplitude *ac* drives also lead to phase homogenization, and to interference effects which survive the passage to the thermodynamic limit. Whether or not this corresponds to the conclusions arrived at on the basis of the hydrodynamic treatment of the CDW dynamics (Sneddon et al., 1988) remains to be seen.

Considerable theoretical activity, generated by the observation of subharmonic interference peaks, led to several fundamentally different proposals on the origin of subharmonic locking. Early analysis indicated that the concept of Devil's staircase behavior tied to inertial effects may be appropriate. This possibility is now considered unlikely; two remaining suggestions, nonsinusoidal pinning potential and internal degrees of freedom, are able to explain a broad variety of interference features and mode locking. Most probably, both are important as there is no *a priori* reason why the pinning potential should be sinusoidal; and there is also a broad variety of independent experimental evidence for the importance of the dynamics of the internal modes of the condensate. The former effect may be more dominant for small specimens where pinning by boundaries such as the surface of the specimens and contacts are important; while the latter may dominate for larger and impure materials where pinning is due to randomly distributed impurities. While the experiments led to several interesting theoretical questions; such as why internal degrees of freedom dynamics mimics many effects which are also the consequence of inertial dynamics, or the concept of minimally stable states; the hope that driven charge density waves can be appropriate model systems for general questions concerning nonlinear dynamics now appears to be remote. This is mainly because both randomly distributed pinning centers (which lead to a broadly defined localization problem) and extended pinning centers (leading to essentially a boundary problem) are, in general, equally important and cannot be easily separated.

Several issues related to the dynamics of charge density waves have not been discussed in this chapter. In particular, the charge density wave dynamics of specimens which display a so-called "switching" behavior (Zettl and Grüner, 1983; Hall et al., 1986) has not been mentioned. The phenomenon arises as the consequence of macroscopic defects which lead to dynamics similar to those observed in coupled Josephson junctions (Inoui and Doniach, 1987). Several observations concerning transitions to chaos have also not been discussed and I refer to a recent review (Zettl and Grüner, 1986), which covers this type of experiment. Mode-locking phenomena also occur in the elastic properties (Bourne et al., 1986), a not too surprising observation given the fact that the CDW can be regarded as a coupled electron-phonon system, and the earlier observations on nonlinear elastic properties.

Several models, with interesting dynamical properties have not been discussed. Among them, the dynamics of Frenkel-Kontorova type models have been studied in detail (Sneddon, 1984a; Coppersmith and Littlewood, 1985a, b), and have displayed many of the features which are the consequence of models where random impurity pinning is explicitly taken into account. Such models also lead to a broad variety of interference effects in the presence of combined *dc* and *ac* electric fields, as discussed earlier.

References

Chapter 1

Heeger, A.J., 1979, in *Highly Conducting One-Dimensional Solids*, edited by J. Devreese et al. (Plenum Press; New York, London).
Kittel, C., 1963, *Quantum Theory of Solids* (John Wiley and Sons, New York).
Schrieffer, J.R., 1964, *Theory of Superconductivity* (W.A. Benjamin, Inc., New York).
Solyom, J., 1979, *Adv. Phys.* **28**, 201.
Emery, V., 1979, in *Highly Conducting One-Dimensional Solids*, edited by J. Devreese et al. (Plenum Press; New York, London).
Tinkham, M., 1975, *Introduction to Superconductivity* (McGraw Hill, New York).
Ziman, I.M., 1964. *Principles of the Theory of Solids*, (Cambridge University Press; Cambridge, London, New York and Melbourne).

Chapter 2

Bechgaard, K. et al., 1980, *Solid State Comm.* **33**, 1119.
Bruesh, H. et al., 1975, *Phys. Rev.* **B12**, 219.
Carneiro, K., 1985, *Electronic Properties of Inorganic Quasi-One-Dimensional Compounds*, edited by P. Monceau (D. Reidel Publ. Co.; Dordrecht, Boston).
Geserich, H.P., 1985, *Electronic Properties of Inorganic Quasi-One-Dimensional Compounds*, edited by P. Monceau (D. Reidel Publ. Co.; Dordrecht, Boston).
Gressier, P. et al., 1984, *Inorg. Chem.* **23**, 1221.
Jacobsen, C.S., D.B. Tanner and K. Bechgaard, 1983, *J. Phys. Colloq.* **46**, 21.
Jerome, D. and H. Schultz, 1982, *Adv. Phys.* **32**, 299.
Kikuchi, K. et al., 1987, *Jap. J. Appl. Phys.* **26**, Suppl. 26-3, 1369.
Kanoda, K. et al., 1988, *Phys. Rev.* **B38**, 39.
Nakamura, T. et al., 1990, *Solid State Comm.* **75**, 583.
Meerchant, A. and J. Rouxel, 1986, in *Crystal Chemistry and Properties of Materials with Quasi-One-Dimensional Structures*, edited by J. Rouxel (D. Reidel Publ. Co.; Dordrecht, Boston), p. 205.

Rouxel, J. and C. Schenker, 1989, in *Charge Density Waves in Solids*, edited by L.P. Gor'kov and G. Grüner, "Modern Problems in Condensed Matter Sciences" (North Holland; Amsterdam, Oxford, New York, Tokyo).
Schlenker, C. and J. Dumas, 1986, in *Crystal Chemistry and Properties of Materials with Quasi-One-Dimensional Structures*, edited by J. Rouxel (D. Reidel Publ. Co.; Dordrecht, Boston).
Travaglini, G. et al., 1981, *Solid State Comm.* **37**, 599.
Whangboo, M.H. and L. Schneemeyer, 1986, *Inorg. Chem.* **25**, 2424.

Chapter 3

Allender, D., J.W. Bray and J. Bardeen, 1974, *Phys. Rev.* **B9**, 119.
Artemenko, S.N. and A.F. Volkov, 1983, *Pisma Zh. Eksp. Tear Fiz.* **17**, 310.
Berthier, C. and P. Segranson, 1987, in *Low Dimensional Conductors and Superconductors*, edited by D. Jerome and L.G. Caron (Plenum Press, New York).
Berthier, C. et al., 1992, in *Nuclear Spectroscopy on CDW Systems*, edited by T. Butz and A. Lerf (Kluwer Academic Publ., Dordrecht, Boston, London).
Comes, R. et al., 1975, *Phys. Stat. Sol.* **B71**, 171.
Dai, Z., G. Slough and R.V. Coleman, *Phys. Rev. Lett.* **66**, 1318 (1991) and references cited therein.
Eagen, C.F., S.A. Werner and R.B. Saillant, 1975, *Phys. Rev.* **B12**, 2036.
Ekino, T. and J. Akimishu, 1987, *Jap. J. Appl. Phys.* **26**, Suppl. 625.
Fournel, A. et al., 1986, *Phys. Rev. Lett.* **57**, 2199.
Frölich, H., 1954, *Proc. Roy. Soc. London Ser.* **A223**, 296.
Grüner, G., 1988, *Rev. Mod. Phys.* **60**, 1129.
Huang, X. and Maki, K. 1989, *Phys. Rev.* **B40**, 2575.
Jánossy, A., C. Berthier and P. Segransan, 1987, *Phys. Scr.* **19B**, 578.
Johnston, D.C., M. Maki and G. Grüner, 1985, *Solid State Comm.* **53**, 5.
Johnston, D.C., 1984, *Phys. Rev. Lett.* **52**, 2049.
Kuper, C.G., 1955, *Proc. R. Soc.* **A227**, 214.
Kwok, R. et al., 1990, *Phys. Rev. Lett.* **65**, 365.
Maki, K. and A. Virosztek., 1990, *J. Magn. Magn. Mat.* **90**, 758.
Monceau, P., 1985, in *Electronic Properties of Inorganic Quasi-One-Dimensional Compounds*, edited by P. Monceau (D. Reidel Publ. Co.; Dordrecht, Boston).
Peierls, R.E., 1955, *Quantum Theory of Solids* (Oxford University Press, New York).
Pouget, J.P. and R. Comes, 1989, in *Charge Density Waves in Solids*, edited by L.P. Gor'kov and G. Grüner (North Holland; Amsterdam, Oxford, New York, Tokyo).
Pouget, J.P., 1987, in *Low Dimensional Conductors and Superconductors*, NATO ASI No. B155, edited by D. Jerome and L.G. Caron (Plenum Press, New York), 1987.
Rice, T.M., P.A. Lee and M.C. Cross, 1979, *Phys. Rev.* **B20**, 1345.
Ross, J.H., Z. Wang and C. Schlichter, 1986, *Phys. Rev. Lett.* **56**, 663.

Schrieffer, J.R., 1964, *Theory of Superconductivity* (Benjamin, Inc.; New York, Amsterdam).
Slough, C.G. et al., 1989, *Phys. Rev.* **B39**, 5496.
Tinkham, M., 1975, *Introduction to Superconductivity* (R.E. Krieger, New York).
Woll, E.J. and W. Kohn, 1962, *Phys. Rev.* **126**, 1693.

Chapter 4

Anderson, P.W., 1990, in *Current Trends in Condensed Matter, Particle Physics, and Cosmology*, edited by J. Pati et al. (World Scientific; Singapore, New Jersey, London, Hong Kong).
Andrieux, A. et al., 1981, *J. Phys. Lett.* **42**, L87.
Bechgaard, K. et al., 1980, *Solid State Comm.* **33**, 1119.
Creuzet, F. et al., 1982, *J. Phys. Lett.* **43**, L755.
Delrieux, J.M. et al., 1986, *Physica* **143D**, 412.
Huang, X. and K. Maki, 1990, *Phys. Rev.* **B42**, 6498.
Kanoda, K. et al., 1988, *Phys. Rev.* **B38**, 39.
Ibid., 1990, **B42**, 8678.
Kikuchi, K. et al., 1987, *Jap. J. Appl. Phys.* **26**, Suppl. 26-3, 1369.
Le, L.P. et al., 1991, *Europhys. Lett.* **15**, 547.
Mortensen, K. et al., 1981, *Phys. Rev. Lett.* **46**, 1234; ibid., 1982, **B25**, 3319.
Nakamura, T. et al., 1990, *Solid State Comm.* **75**, 583.
Overhauser, A.W., 1960, *Phys. Rev. Lett.* **4**, 462.
Overhauser, A.W., 1962, *Phys. Rev.* **128**, 1437.
Takahashi, T. et al., 1986, *J. Phys. Soc. Japan* **55**, 1364.
Yamaji, J. 1982, *J. Phys. Soc. Japan* **5**, 2787; ibid., 1983, **52**, 1369.

Chapter 5

Aronowitz, J.A., P. Goldbart and G. Mozurkewich, 1990, *Phys. Rev. Lett.* **64**, 2799.
Degiorgi, L. and G. Grüner 1992, *de Physique* **2**, 253
Dietrich, W., 1976, *Adv. Phys.* **25**, 615.
Ginzburg, V.L., 1960, *S. Phys. Sol. State* **2**, 1824.
Girault, S. et al., 1988, *Phys. Rev.* **B38**, 7980.
Gorshunov, B.P. et al., (to be published).
Hauser, M.R. et al., 1991, *Phys. Rev.* **B43**, 8105.
Horowitz, B. et al., 1975, *Phys. Rev.* **B12**, 3174.
Johnston, D.C., M. Maki and G. Grüner, 1985, *Solid State Comm.* **53**, 5.
Kwok, R. and S.E. Brown, 1989, *Phys. Rev. Lett.* **63**, 895.
Kwok, R., S.E. Brown and G. Grüner, 1990, *Phys. Rev. Lett.* **65**, 365.
Landau, L.D. and E.M. Lifschitz, 1959, *Statistical Physics* (Pergamon Press, London).
Lee, P.A., T.M. Rice and P.W. Anderson, 1973, *Phys. Rev. Lett.* **31**, 462.

Pouget, J.P., 1989, in *Low Dimensional Electronic Properties of Molybdenum Bronzes and Oxides*, edited by C. Schlenker (Kluver Academic Publ.).
Pouget, J.P. and R. Comes, 1989, in *Charge Density Waves in Solids*, edited by L.P. Gor'kov and G. Grüner (North Holland; Amsterdam, Oxford, New York, Tokyo).
Scalapino, D.J., Y. Imry and P. Pincus, 1975, *Phys. Rev.* **B11**, 2042.
Schulz, H.J., 1987, in *Low Dimensional Conductors and Superconductors*, edited by E.D. Jerome and H.G. Caron (NATO ASI, Plenum Press, New York).
Soub, K. et al., 1976, *Phys. Lett.* **A56**, 302.

Chapter 6

Anderson, P.W., 1984, *Basic Notions in Condensed Matter Physics* (Benjamin-Cummings, Redwood City, CA).
Barisic, S., 1985, *Electronic Properties of Materials with Quasi-One-Dimensional Structures*, edited by P. Monceau (D. Reidel Publ. Co., Dordrecht, Boston).
Boriak, S. and A.W. Overhauser, 1977, *Phys. Rev.* **B16**, 5206.
Brazovskii, S. and I.E. Dzyaloshihskii, 1976, *Sov. Phys. JETP* **44**, 1233.
Carlson, R.V. and A.M. Goldman, 1973, *Phys. Rev. Lett.* **32**, 880; ibid., 1975, **34**, 11.
Escribe-Filippini, C. et al., 1987, *Synth. Met.* **19**, 931.
Fukuyama, H. and P.A. Lee, 1978, *Phys. Rev.* **B17**, 535.
Kittel, C., 1963, *Quantum Theory of Solids* (John Wiley and Sons, New York).
Kurihara, Y., 1980, *J. Phys. Soc. Japan* **49**, 852.
Le, L.P. et al., 1993, *Phys. Rev. B.* (in press).
Lee, P.A. and H. Fukuyama, 1978, *Phys. Rev.* **B17**, 542.
Lee, P.A., T.M. Rice and P.W. Anderson, 1974, *Solid State Comm.* **14**, 703.
Littlewood, P.B. and C.M. Varma, 1982, *Phys. Rev.* **B26**, 4883.
Maki, K. and G. Grüner, 1991, *Phys. Rev. Lett.* **66**, 789.
Maki, K. and A. Virosztek, 1990, *J. Magn. Magn. Mat.* **90**, 758.
Maki, K. and A. Virosztek, 1992, *Phys. Rev. Lett.* **69**, 3265.
Nakane, Y. and S. Takada, 1985, *J. Phys. Soc. Japan* **54**, 977.
Pouget, J.P. et al., 1992, *Phys. Rev. Lett.* **68**, 2374.
Pouget, P. and R. Comes, 1989, in *Charge Density Waves in Solids*, edited by L.P. Gor'kov and G. Grüner, "Modern Problems in Condensed Matter Sciences" (North Holland; Amsterdam, Oxford, New York, Tokyo).
Psaltakis, G.C., 1987, *Solid State Comm.* **51**, 535.
Schrieffer, J.R., X.G. Wen and S.C. Zhang, 1989, *Phys. Rev.* **B39**, 11663.
Schultz, H.J., 1977, Thesis, University of Hamburg (unpublished).
Torrance, J.B. et al., 1982, *Phys. Rev. Lett.* **49**, 881.
Travaglini, G. et al., 1983, *Solid State Comm.* **45**, 289.
Virosztek, A. and K. Maki, 1993, *Phys. Rev. B* (in press).

Chapter 7

Bak, Per and V.L. Pokrovsky, 1981, *Phys. Rev. Lett.* **47**, 958.
Bishop, A.R., J.A. Krumansl and S.E. Trullinger, 1990, *Physica* **ID**.
Brazovskii, S., 1989, in *Charge Density Waves in Solids*, edited by L.P. Gor'kov and G. Grüner "Modern Problems in Condensed Matter Sciences" (North Holland; Amsterdam, Oxford, New York, Tokyo).
Horowitz, B., 1986, in *Solitons*, eds: S. Trullinger et al., "Modern Problems in Condensed Matter Sciences" (North Holland; Amsterdam, Oxford, New York, Tokyo).
Lee, P.A., T.M. Rice and P.W. Anderson, 1979, *Solid State Comm.* **14**, 703.
Maki, K. in *Electronic Properties of Inorganic Quasi-One-Dimensional Compounds*, edited by P. Monceau (D. Reidel Publ. Co., Dordrecht, Boston).
Pouget, P. and R. Comes, 1989, in *Charge Density Waves in Solids*, edited by L.P. Gor'kov and G. Grüner (North Holland; Amsterdam, Oxford, New York, Tokyo).
Rice, M.J. et al., 1976, *Phys. Rev. Lett.* **36**, 342.
Su, W.P. and J.R. Schrieffer, 1981, *Phys. Rev. Lett.* **46**, 738.

Chapter 8

Biljakovich, K. et al., 1981, *Phys. Rev. Lett.* **57**, 1907.
Biljakovich, K., J.C. Lasjammias and P. Monceau, 1989, *Synth. Met.* **29**, F289.
Brown, S.E. et al., 1988, *Phys. Rev.* **B37**, 6551.
DeLand, S.M., G. Mozurkewich and L.D. Chapman, 1991, *Phys. Rev. Lett.* **66**, 2026.
Efetov, K.B. and A.I. Larkin, 1977, *Zh. Eksp. Tear. Fiz.* **72**, 7350.
Forro, L. et al., 1983, *Phys. Rev.* **B27**, 7600.
Fukuyama, H. and P.A. Lee, 1978, *Phys. Rev.* **B17**, 535.
Girault, S. et al., 1988, *Phys. Rev.* **B38**, 7980.
Imry, I. and S.K. Ma, 1975, *Phys. Rev. Lett.* **35**, 1399.
Lee, P.A. and T.M. Rice, 1979, *Phys. Rev.* **B19**, 3970.
Sham, L.J. and B.R. Patton, 1976, *Phys. Rev.* **B13**, 2151.
Sweetland, E. et al., 1990, *Phys. Rev. Lett.* **75**, 3165.
Tamagai, T., K. Tsutsumi and S. Kagoshima, 1987, *Synth. Met.* **19**, 923.
Tüttő, I. and A. Zawadowski, 1985, *Phys. Rev. Lett.* **B32**, 2449.

Chapter 9

Degiorgi, L. et al., 1991, *Phys. Rev.* **B44**, 7808.
Donovan, S. et al. *Phys. Rev.* (to be published).
Frölich, H., 1954, *Proc. R. Soc. London Ser.* **A223**, 296.
Grüner, G., 1988, *Mod. Phys.* **60**, 1129 (and references).
Lee, P.A., T.M. Rice and P.W. Anderson, 1974, *Solid State Comm.* **14**
Kim, T.W. et al., 1991, *Phys. Rev.* **B43**, 6315.
Littlewood, P.B., 1987, *Phys. Rev.* **B36**, 3108.
Ngai, K.L., 1979, *Comments Solid State Phys.* **9**, 127.

Pippard, A.B., 1953, *Proc. R. Soc. London Ser.* **A216**, 547.
Quinlivan, D. et al., 1990, *Phys. Rev. Lett.* **65**, 1816.
Rice, M.J., 1976, *Phys. Rev. Lett.* **37**, 36.
Tüttő, I. and A. Zawadowski, 1985, *Phys. Rev. Lett.* **32**, 2449.
Wonneberger, W., 1991, *Synth. Met.* **41**, 3793 (and references cited therein).
Wooten, F., 1972, *Optical Properties of Solids* (Academic Press).
Zawadowski, A., 1987, *Synth. Met.* **29**, F469.

Chapter 10

Barone, A. and G. Paterno, 1982, *Physics and Applications of the Josephson Effect* (John Wiley and Sons, New York).
Beyermann, W.P., L. Mihály and G. Grüner, 1986, *Phys. Rev. Lett.* **56**, 1489.
Fleming, R.M. et al., 1978, *Phys. Rev.* **B18**, 5560.
Gor'kov, L.P., 1977, *JETP Lett.* **25**, 358.
Grüner, G., 1988, *Rev. Mod. Phys.* **60**, 1129.
Grüner, G., 1994, *Rev. Mod. Phys.* January issue.
Grüner, G., A. Zawadowski and P.M. Chaikin, 1981, *Phys. Rev. Lett.* **46**, 511.
Grüner, G., W.G. Clark and A.M. Portis, 1981, *Phys. Rev.* **B24**, 3641.
Grüner, G. and A. Zettl, 1985, *Physics Rep.* **119**, 117.
Littlewood, P.B., 1989, in *Charge Density Waves in Solids*, edited by L.P. Gor'kov and G. Grüner (North Holland; Amsterdam, Oxford, New York, Tokyo).
Mihály, G. et al., 1991, *Phys. Rev. Lett.* **66**, 1806.
Mihály, G. et al., 1988, *Phys. Rev.* **B37**, 1097.
Monceau, P. et al., 1976, *Phys. Rev. Lett.* **37**, 6902.
Myers, C.R. and J.P. Sethna, 1993, *Phys. Rev.* **B47**, 11171.
Narayan, O. and D.S. Fisher, 1992, *Phys. Rev.* **B46**, 11520.
Nomura, K., 1986, *Physica*, **B143**, 117.
Ross, J.H., Z. Wang and C. Slichter, 1986, *Phys. Rev. Lett.* **56**, 663.
Segranson, P. et al., 1986, *Phys. Rev. Lett.* **56**, 1854.
Tomic, S. et al., 1989, *Phys. Rev. Lett.* **62**, 462.
Zettl, A. and G. Grüner, 1982, *Phys. Rev.* **B26**, 2298.

Chapter 11

Alstrom, P., M.H. Jensen and M.T. Levinsen, 1984, *Phys. Rev. Lett.* **A103**, 171.
Azbel, M.Ya. and Per Bak, 1984, *Phys. Rev.* **B30**, 3722.
Bak, Per, 1983, in *Proc. Int. Symp. on Nonlinear Transport in Inorganic-Quasi-One-Dimensional Conductors* (Sapporo, Japan) (unpublished).
Bak, Per, T. Bohr, M.H. Jensen and R.V. Christianson, 1984, *Solid State Comm.* **51**, 231.
Bardeen, J., E. Ben-Jacob, A. Zettl and G. Grüner, 1982, *Phys. Rev. Lett.* **49**, 493.
Beauchene, P., J. Dumas, A. Jánossy, J. Marcus and C. Schlenker, 1986, *Physica* **B143**, 126.

Bhattacharya, S., J.P. Stokes, M. Robbins and R.A. Klemm, 1985, *Phys. Rev. Lett.* **54** 2453.
Bhattacharya, S., J.P. Stokes, M.J. Higgins and R.A. Klemm, 1987, *Phys. Rev. Lett.* **59**, 1849.
Borodin, D.V., F.Ya. Nad', Ya.S. Savitskaya and S.V. Zaitsev-Zotov, 1986, *Physica* **B143**, 73.
Bourne, L.C., M.S. Sherwin and A. Zettl, 1986, *Phys. Rev. Lett.* **56**, 1952.
Brown, S.E. and G. Grüner, 1985, *Phys. Rev.* **B31**, 8302.
Brown, S.E. and L. Mihály, 1985, *Phys. Rev. Lett.* **55**, 742.
Brown, S.E., G. Mozurkewich and G. Grüner, 1984, *Phys. Rev. Lett.* **52**, 2277.
Brown, S.E., G. Mozurkewich and G. Grüner, 1985a, *Solid State Comm.* **54**, 23.
Brown, S.E., A. Jánossy and G. Grüner, 1985b, *Phys. Rev.* **B31**, 6869.
Brown, S.E., G. Grüner and L. Mihály, 1986a, *Solid State Comm.* **57**, 165.
Brown, S.E., L. Mihály and G. Grüner, 1986b, *Physica* **D23**, 169.
Clark, T.D. and P.E. Lindelof, 1976, *Phys. Rev. Lett.* **37**, 368.
Coppersmith, S.N. and P.B. Littlewood, 1985a, *Phys. Rev.* **B31**, 4049.
Coppersmith, S.N. and P.B. Littlewood, 1985b, in *Proc. Int. Conf. on Charge Density Waves in Solids, Lecture Notes in Physics*, Vol. 217, edited by J. Solyom and Gy. Hutiray, Springer Verlag, Berlin, p. 236.
Coppersmith, S.N. and P.B. Littlewood, 1986, *Phys. Rev. Lett.* **57**, 1927.
Fack, H. and V. Kose, 1971, *J. Appl. Phys.* **42**, 320.
Fisher, D., 1985, *Phys. Rev.* **B31**, 1396.
Fleming, R.M., 1983, *Solid State Comm.* **43**, 169.
Fleming, R.M. and Grimes, C.C., 1979, *Phys. Rev. Lett.* **42**, 1423.
Fleming, R.M. and L.F. Schneemeyer, 1986, *Phys. Rev.* **B33**, 2930.
Fleming, R.M., L.F. Schneemeyer and R.J. Cava, 1985, *Phys. Rev.* **B31**, 1181.
Fukuyama, H. and P.A. Lee, 1978, *Phys. Rev.* **B17**, 535.
Gill, J.C., 1982, *Solid State Comm.* **44**, 1041.
Grüner, G., 1988, *Rev. Mod. Phys.* **60**, 1129.
Hall, R.P. and A. Zettl, 1984, *Phys. Rev.* **B30**, 2279.
Hall, R.P., M.F. Hundley and A. Zettl, 1986, *Physica* **B143**, 152.
Inoue, M. and S. Doniach, 1987, *Phys. Rev.* **B33**, 6244.
Jing, T.W. and N.P. Ong, 1986, *Phys. Rev.* **B33**, 5841.
Latyshev, Yu.I., V.E. Minakova, Ya.S. Santikaya and V.V. Frolov, 1986, *Physica* **B143**, 155.
Latyshev, Yu.I., V.E. Minakova and Ya.A. Zhanov, 1987, *Pis'ma v Zs. Eksp. and Teor. Fiz.* **46**, 31.
Lee, P.A. and T.M. Rice, 1979, *Phys. Rev.* **B19**, 3970.
Lindelof, P.E., 1981, *Rep. Prog. Phys.* **44**, 949.
Link, G.L. and G. Mozurkewich, 1988, *Solid State Comm.* **65**, 15.
Littlewood, P.B., 1986, *Phys. Rev.* **B33**, 6694.
Lyding, J.W., J.S. Hubacek, G. Gammic and R.E. Thorne, 1986, *Phys. Rev.* **B33**, 4341.
Maeda, A.N., Naito and S. Tanaka, 1983, *Solid State Comm.* **47** 1001.
Maeda, A.N., Naito and S. Tanaka, 1985, *J. Phys. Soc. Japan* **54**, 1912.
Matsukawa, H., 1987, *J. Phys. Soc. Japan* **56**, 1522.

Monceau, P., J. Richard and M. Renard, 1980, *Phys. Rev. Lett.* **45**, 43.
Monceau, P., M. Renard, J. Richard, M.C. Saint-Lager, H. Salva and Z.Z. Wang, 1983, *Phys. Rev.* **B28**, 1646.
Monceau, P., M. Renard, J. Richard and M.C. Saint-Lager, 1986, *Physica* **B143**, 64.
Mozurkewich, G. and G. Grüner, 1983, *Phys. Rev. Lett.* **51**, 2206.
Mozurkewich, G., M. Maki and G. Grüner, 1983, *Solid State Comm.* **48**, 453.
Ong, N.P. and K. Maki, 1985, *Phys. Rev.* **B32**, 6582.
Ong, N.P., G. Verma and K. Maki, 1984a, *Phys. Rev. Lett.* **52**, 663.
Ong, N.P., C.B. Kalem and J.C. Eckert, 1984b, *Phys. Rev.* **B30**, 2902.
Parrilla, P. and A. Zettl, 1985, *Phys. Rev.* **B32**, 8427.
Renne, M.J. and D. Poulder, 1974, *Rev. Phys. Appl.* **9**, 25.
Richard, J., P. Monceau, H. Papoulas and M. Renard, 1982, *J. Phys.* **C15**, 7157.
Segransan, P., A. Jánossy, C. Berthier, J. Marcus and P. Boutaud, 1986, *Phys. Rev. Lett.* **56**, 1954.
Shapiro, S., 1963, *Phys. Rev. Lett.* **11**, 80.
Sherwin, M.S., and A. Zettl, 1985, *Phys. Rev.* **B32**, 5536.
Sneddon, L., 1984, *Phys. Rev. Lett.* **52**, 65.
Sneddon, L., M. Cross and D. Fisher, 1982, *Phys. Rev. Lett.* **49**, 292.
Thorne, R.E., W.G. Lyons, J.M. Miller, J.W. Lyding and J.R. Tucker, 1986a, *Phys. Rev.* **B34**, 5988.
Thorne, R.E., J.R. Tucker, J. Bardeen, S.E. Brown and G. Grüner, 1986b, *Phys. Rev.* **B33**, 7342.
Thorne, R.E., W.G. Lyons, J.W. Lyding, J.R. Tucker and J. Bardeen, 1987a, *Phys. Rev.* **B35**, 6360.
Thorne, R.E., J.R. Tucker and J. Bardeen, 1987b, *Phys. Rev. Lett.* **58**, 828.
Thorne, R.E., J.S. Hubacek, W.G. Lyons, J.W. Lyding and J.R. Tucker, 1988, *Phys. Rev.* **B37**, 10055.
Tua, P.F. and J. Ruvalds, 1985, *Solid State Comm.* **54**, 471.
Tua, P.F. and Z. Zawadowski, 1984, *Solid State Comm.* **49**, 19.
Verma, G. and N.P. Ong, 1984, *Phys. Rev.* **B30**, 2928.
Waldram, J.R. and R.H. Wu, 1982, *J. Low Temp. Phys.* **47**, 363.
Weger, M., G. Grüner and W.G. Clark, 1980, *Solid State Comm.* **35**, 243.
Weger, M., G. Grüner and W.G. Clark, 1982, *Solid State Comm.* **44**, 1179.
Yeh, W.J., Da-Run He and Y.H. Kao, 1984, *Phys. Rev. Lett.* **52**, 480.
Zettl, A. and G. Grüner, 1983, *Solid State Comm.* **46**, 501.
Zettl, A. and G. Grüner, 1984, *Phys. Rev.* **B29**, 755.
Zettl, A. and G. Grüner, 1986, *Comments Cond. Matter Phys.* **12**, 265.
Zettl, A., M.B. Kaiser and G. Grüner, 1985, *Solid State Comm.* **53**, 649.

Appendix
Some books, conference proceedings, and review papers

Books

Devreese, J.T. et al., editors, 1979, *Highly Conducting One-Dimensional Solids* (Plenum Press, New York).

Gor'kov, L.P. and G. Grüner, editors, 1989, *Charge Density Waves in Solids*, in *Modern Problems in Condensed Matter Sciences*, edited by V.M. Agranovich and A.A. Maradudin (North Holland; Amsterdam, Oxford, New York, Tokyo).

Kagoshima, S., H. Nagasawa and T. Sambongi, 1988, *One-Dimensional Conductors*, Spring Series in Solid-State Sciences 72 (Springer-Verlag; Berlin).

P. Monceau, editor, 1985, *Electronic Properties of Inorganic Quasi-One-Dimensional Compounds* (D. Reidel Publ. Co., Dordrecht, Boston).

J. Rouxel, editor, 1986, *Crystal Chemistry and Properties of Materials with Quasi-One-Dimensional Structures* (D. Reidel Publ. Co.; Dordrecht, Boston).

Conference Proceedings

Alcacer, L., editor, 1979, *The Physics and Chemistry of Low-Dimensional Solids* (D. Reidel Publ. Co., Dordrecht, Boston).

Barisic, S. et al., editors, 1978, "Quasi-One-Dimensional Conductors", *Lecture Notes in Physics* (Springer-Verlag; Berlin).

Bernasconi, J., editor, 1981, *Physics in One Dimension* (Springer-Verlag, Berlin, Heidelberg).

Gutiray, Gy. and J. Solyom, editors, 1985, "Charge Density Waves in Solids," *Lecture Notes in Physics* (Springer-Verlag, Berlin).

Jerome, D. and L.G. Caron, editors, 1987, *Low-Dimensional Conductors and Superconductors* (NATO ASI Plenum Press, New York).

Keller, H.J., editor, 1974, *Low-Dimensional Cooperative Phenomena* (NATO ASI Plenum Press, New York).

"Yamada Conference on One-Dimensional Conductors, Lake Kawasaki," 1986, *Physica B. International Conference on Synthetic Low-Dimensional Conductors and Superconductors*, *J. Phys.* (France) 44-C3.

Review Papers

Berlinski, A.J., 1976, "One-dimensional metals and charge density wave effects," *Contemp. Phys.* **17**, 331.

Dieterich, W., 1976, "Ginzburg-Landau, theory of phase transitions in pseudo one-dimensional systems," *Adv. Phys.* **25**, 615.

Grüner, G., 1988, "The dynamics of charge density waves," *Rev. of Mod. Phys.* **60**, 1129.

Grüner. G., 1994, "The dynamics of spin density waves," *Rev. of Mod. Phys.* January issue.

Grüner, G. and K. Maki, 1991, "The dynamics of spin density waves," *Comm. in Cond. Mat. Phys.* **15**, 145.

Grüner, G. and A. Zettl, 1985, "Charge density wave conduction: a novel collective transport phenomenon in solids," *Phys. Rep.* **119**, 117.

Horowitz, B., 1986, "Solitons in charge and spin density wave systems in 'solitons'," edited by S.E. Trullinger et al., *Modern Problems in Condensed Matter Sciences Series*, **17** (North Holland, Amsterdam).

Jerome, D. and H. Schulz, 1982, "Organic conductors and superconductors," *Adv. Phys.* **31**, 299.

Krive, I.V., A.S. Rozhovskii and I.O. Kulik, 1986, "Mechanism of nonlinear conductivity and electrodynamics of one-dimensional conductors in the Peierls insulator state," *Soviet J. of Low Temp. Phys.* **12**, 635.

Solyom, J., 1979, "The Fermi gas model of one-dimensional conductors," *Adv. Phys.* **28**, 201.

Toombs, G.A., 1978, "Quasi-one-dimensional conductors," *Phys. Rep.* **40**, 1981.

Index

Printed in the United States
by Baker & Taylor Publisher Services

Printed in the United States
by Baker & Taylor Publisher Services